DEBUNKING HUMAN EVOLUTION TAUGHT IN OUR PUBLIC SCHOOLS

A Guidebook for Christian Students, Parents, and Pastors

Daniel A. Biddle, Ph.D., David A. Bisbee, & Jerry Bergman, Ph.D.

Copyright © 2016 by Genesis Apologetics, Inc.
E-mail: staff@genesisapologetics.com

 GENESIS apologetics

http://www.genesisapologetics.com: A 501(c)(3) ministry equipping youth pastors, parents, and students with Biblical answers for evolutionary teaching in public schools.

DEBUNKING HUMAN EVOLUTION TAUGHT IN OUR PUBLIC SCHOOLS
A Guidebook for Christian Students, Parents, and Pastors
by Daniel A. Biddle, Ph.D., David A. Bisbee, & Jerry Bergman, Ph.D.
Printed in the United States of America

ISBN: 978-1-944918-05-7

Library of Congress Control Number: 2016932943

Dedication

To my wife, Jenny, who supports me in this work. To my children Makaela, Alyssa, Matthew, and Amanda, and to your children and your children's children for a hundred generations—this book is for all of you. To Dave Bisbee and Mark Johnston, who planted the seeds and the passion for this work.

We would like to acknowledge Answers in Genesis (*www.answersingenesis.com*), the Institute for Creation Research (*www.icr.org*), and Creation Ministries International (*www.creation.com*). Much of the content herein has been drawn from (and is meant to be in alignment with) these Biblical Creation ministries.

Guard what has been entrusted to you, avoiding worldly and empty chatter and the opposing arguments of what is falsely called "knowledge"—which some have professed and thus gone astray from the faith. Grace be with you.
— 1 Tim. 6:20–21

"This is the Lord's doing; it is marvelous in our eyes."
— Psalm 118:23

CONTENTS

ABOUT THE AUTHORS

Dr. Jerry Bergman has five Master's degrees in the science, health, psychology, and biology fields, a Ph.D. in human biology from Columbia Pacific University, and a Ph.D. in measurement and evaluation from Wayne State University. Jerry has taught biology, genetics, chemistry, biochemistry, anthropology, geology, and microbiology at Northwest State Community College in Archbold Ohio for over thirty years. He is currently an Adjunct Associate Professor at the University of Toledo Medical School. He has over one thousand publications in twelve languages and thirty-two books and monographs, including *Vestigial Organs Are Fully Functional*; *Slaughter of the Dissidents: The Shocking Truth About Killing the Careers of Darwin Doubters*; *The Dark Side of Darwin*; *Hitler and the Nazis Darwinian Worldview: How the Nazis Eugenic Crusade for a Superior Race Caused the Greatest Holocaust in World History* and *The Darwin Effect. Its influence on Nazism, Eugenics, Racism, Communism, Capitalism & Sexism*. Professor Bergman has also taught at the Medical College of Ohio as a research associate in the Department of Experimental Pathology, at the University of Toledo, and at Bowling Green State University. Jerry is also a member of MENSA, a Fellow of the American Scientific Association, a member of The National Association for the Advancement of Science, and member of many other professional associations. He is listed in Who's Who in America, Who's Who in the Midwest and in Who's Who in Science and Religion.

Dr. Daniel A. Biddle is president of Genesis Apologetics, Inc. a 501(c)(3) organization dedicated to equipping youth pastors, parents, and students with Biblical answers for evolutionary teaching in public schools. Daniel has trained thousands of students in Biblical creation and evolution and is the author of several Creation-related publications. Daniel has a Ph.D. in Organizational Psychology from Alliant University in San Francisco, California, an M.A. in Organizational Psychology from Alliant, and a B.S. in Organizational Behavior from the University of San Francisco. Daniel has worked as an expert consultant and/or witness in over one hundred state and federal cases in the areas of research methodologies and analysis.

David A. Bisbee is the Vice President of Genesis Apologetics, Inc., a non-profit 501(c)(3) organization that equips Christian students attending public schools and their parents with faith-building materials that reaffirm a Biblical creation worldview. Genesis Apologetics is committed to providing Christian families with Biblically- and scientifically-based answers to the evolutionary theory that students are taught in and public schools. Mr. Bisbee's professional experience includes over twenty-five years in the field of energy efficiency. For the last fourteen years, he has been in charge of a research program which tests energy efficiency technologies in real-world environments. Mr. Bisbee has presented the results of these projects through numerous published reports and educational seminars throughout the United States. Dave's ministry experience includes over ten years teaching Sunday school classes and creation science presentations.

Dr. Jeffrey Tomkins has a Ph.D. in Genetics from Clemson University, a M.S. from the University of Idaho, and a B.S. from Washington State University. He was on the Faculty in the Department of Genetics and Biochemistry, Clemson University, for a decade, where he published fifty-seven secular research papers in peer-reviewed scientific journals and seven book chapters on genetics, genomics, proteomics, and physiology. For the past several years, Dr. Tomkins has been a Research Scientist at the Institute for Creation Research and an independent investigator publishing ten peer-reviewed creation science journal papers, numerous semi-technical articles, and two books including *The Design and Complexity of the Cell*.

INTRODUCTION

In science textbooks, the idea that human beings evolved from non-human, ape-like ancestors is treated as a plain, factual truth. This should not be surprising, as it actually stems from a larger tale that is being sold as science: that nature is all there is. Therefore, that story of "natural history" cannot include any sort of creation. If someone believes that nature is all that exists, then life arose without a Creator. This is the necessary conclusion to that story, in spite of the other evidence that is available.

The idea that there is nothing besides nature is known as naturalism and goes at least all the way back to the Greeks. Evolution is not a new idea; it has been held by proponents of naturalism for as long as naturalism has been a philosophy. Lucretius, for instance, writing in the century before Christ, authored *On the Nature of Things*, which included his philosophy on the evolution of animals out of the earth and of man out of the animals. No matter what the evidence shows, if your presumption is that nature is all that there is, your conclusion will be something a lot like evolution.

This type of thinking has led many otherwise excellent scientists astray. While it is true that humans and many apes share some remarkable similarities, the significant differences between them have often been papered over in order to give the illusion of continuity.

Because the idea of evolution has become so entrenched within biology, many students are under the impression that it must be an incontrovertible fact. The textbooks leave no room for discussion on this point. According to the textbooks, the idea that humans had ape-like ancestors is just as certain as Newton's laws of physics.

The fact that these ideas are based on a naturalistic philosophy is never presented to students. Since these "facts" are included in a textbook, students presume that they must be based on rigorous investigation. Therefore, Christian students feel that they must find some way to incorporate evolution into their Christian faith in order to live in a scientific age. Indeed, there is an entire industry of "theistic evolution" publications to allow them do just that.

But the real problem is not that the evidence points to evolution, but that the naturalistic philosophy is accepted as fact by the general population. Is it possible that Christianity actually provides a better way of interpreting the evidence? This book, *Debunking Human Evolution Taught in Our Public Schools*, argues that the Bible offers a better framework to understand the differences between humans and apes, that the Bible better explains the genetic diversity of humans, and that the evidence linking humans and apes has been exaggerated (and sometimes even fabricated) repeatedly in the history of science, and that the evidence continues to be exaggerated to this day.

The Bible has a long history of being strenuously challenged and then being found to be true despite its critics' claims. This is especially true in biology. Experimental biology was born from a strong belief in God's creation. One of biology's first experiments was performed by Francesco Redi, which showed that flies were born from other flies, not from rotting meat. He knew this was true based on the Bible's notion that things are born *according to their kind*. From that idea he constructed the first biology experiment to prove the falsity of spontaneous generation. Most people are unaware that Gregor Mendel founded genetics based on the creation research of Karl Friedrich von Gärtner, and he used that research to argue against the theories of evolution that were popular in his day. Mendel was aiming to provide a theoretical foundation for Gärtner's experiments that showed fixed boundaries to any evolutionary change.

Many people think that the Bible is opposed to the modern biological sciences. As you can see, the very existence of experimental biology and genetics owe their origins to the Bible and Biblical thinking.

If the Bible is true then God created us supernaturally, a reason exists for our creation, and we will eventually answer to God for every decision. On the other hand, if public school textbooks are correct and natural processes made us, we have no lasting purpose, and we will not be held accountable to a Creator after this life. According to Biblical creation, God made Adam and Eve only about six thousand years ago, and all human varieties—living and extinct—descend from the original couple. According to evolution, death of "less fit" creatures transformed populations of ape-like creatures into all human varieties over millions of years. We cannot rewind time to view firsthand the creation of mankind, but we have enormous amounts of evidence supporting creation. Unfortunately, our state textbooks either fail to mention this evidence or explain it away with stories that strictly support naturalistic evolution.

One "proof" that supports biblical creation is that *we see today exactly what the Bible described* in Genesis 1: apes reproducing *after their kinds*, and humans, who are made in the image of God, are very different than animals, likewise reproducing after our kind. The "proof" that supposedly supports evolution comes from human interpretations of fossils and genetics.

Secular textbooks show only arguments *for* interpreting certain fossils, as illustrating a transition from ape to human, and overemphasize similarities between ape and human DNA. These textbooks don't reveal the fact that even some evolutionary scientists use science to refute arguments for evolution. In their zeal to present young readers with a believable-sounding evolutionary history, they don't relate the swirling controversies and confusion that reveals how weak the evidence is for evolutionary views.

For example, just reading through a sixth grade World History textbook might lead the reader to believe that there are thousands of examples of humanlike beings leading up to modern humans. But according to Ian Tattersall, the Director of the American Museum of Natural History, "You could fit it all into the back of a pickup truck if you didn't mind how much you jumbled everything up."[1]

Their DNA analyses claims are equally weak, with no more similarities than necessary to build creatures that eat similar foods and reproduce in similar ways. Apes reproduce apes and humans reproduce humans. No fossil or DNA sequence refutes this.[2]

This book centers around that question. It reveals some of what textbooks have claimed, and it dares to show glimpses of the overwhelmingly powerful case that apes and humans do not share a common ancestry. The latest science reveals how their DNA appears locked in unique but flexible patterns that the Creator intended from the very beginning.

To organize our exploration of human beginnings, this book contrasts the Biblical and evolutionary viewpoints in four sections. The first section will point out the main differences between apes (and other primates) and humans. The second section will summarize human evolution as represented in textbooks over the last 150 years. In the third section we will evaluate the standard line-up of ape-to-human fossil forms found in most school textbooks. You'll learn facts, that your teachers may not even know, about four key fossil forms, each with a fancy name. In the last section we discuss human "races," human and ape genetic similarity, and new genetics discoveries that clearly support recent Creation. Throughout this journey, we will answer questions like:

- What difference does our belief in human origins make in our daily lives?
- Can't we just believe whatever we want and live our lives as good people?

- How does believing in evolution impact my life?
- Why does believing in biblical creation matter so much?

These are important topics because whether or not you were created matters! Think about it—if you believe that humans evolved from apes, then why not just live like you want to live? Without a God, there is no "good," no "evil," and no basic moral laws like the Ten Commandments to guide or govern your life. In this view, there would be no afterlife, no judgment, and no accountability after you die! However, if we believe in a God who made us on purpose, we have purpose, meaning, significance, and accountability in this life, and a hope for everlasting life. Knowing where we came from gives us a firm foundation for daily decisions and even everlasting decisions. This is not just a side issue. It impacts every area of our lives.

CHAPTER 1

Differences between Apes and Humans

Daniel A. Biddle, Ph.D.

WHAT DOES THE BIBLE SAY ABOUT APES AND HUMANS?

In Genesis 1, the very first chapter of the first book in the Bible says that God created land-dwelling creatures including apes on the Sixth Day of Creation. Then He created humans to be able to know the mind of God. Jesus said in Mark 10:6 that humans were created as males and females *at the beginning of Creation* (referring to Genesis 1), not as the result of an evolutionary or any other long-age process. Further, God set a rule in His Creation that apes and humans would only be able to reproduce *after their kind*. If this is true, what would we expect to see? Today, ape mothers always give birth to ape babies, and human mothers to human babies. Do fossils or DNA analyses show that ancient creatures reproduced between kinds instead of according to kinds? Keep studying to find out.

Figure 1. Apes and Humans Today. Millions of apes have really existed on Earth, and billions of people as well. But science reveals that supposedly transitional forms between apes and humans exist only in man-made illustrations.

As Figure 1 indicates, the Bible is correct that many ape-like creatures, each reproducing after their kind, *stay like their kind*, and the same is true for humans. While there is great variability within each primate form (gibbons, gorillas, monkeys, chimpanzees, etc.) there is not a single "in-between" species alive today. No creature looks half ape and half human. While many evolutionists label some fossil forms only part-human, inevitably other evolutionists disagree with them! Details in this book reveal that these same fossils—controversial among evolutionary scientists—easily fit one of four Bible-friendly categories. Two of those categories are "human" and "ape." We'll mention the other two categories later, but for now we know that none of evolution's expected in-between creatures live today.

FIVE BIBLICAL DIFFERENCES BETWEEN HUMANS AND ANIMALS

Humans were the *only* creation that God made in His *own image*: "Then God said, 'Let Us³ make man in Our image, according to Our likeness; and let them rule over the fish of the sea and over the birds of the sky and over the cattle and over all the earth, and over every creeping thing that creeps on the earth.' God created man in His own image, in the image of God He created him; male and female He created them" (Genesis 1:26–27). By God creating man in His image, God made humans with body, soul, and mind. He also gave humans the ability to reason, the power of a will, and the capacity for emotion. This "image of God" includes five unique aspects.

First, Adam was initially made, like God, moral—knowing wrong from right. Adam was given a conscience, embedded in the fiber of his being to tell him how to manage the world and take dominion in a way that honored God's initially perfect Creation.

Second, being created in the image of God also means that we, like God, can *speak*. Unlike all other animals, humans express abstract ideas using speech that no other animal can because we were made in God's image. The English language contains over 1 million words, and we can speak all of them. Apes cannot speak *any* of them. They do not have a speech "program" installed in their brains. They can communicate according to their instincts, but they cannot speak and write using creative, complex language. Speaking and creating go hand-in-hand with being created in the image of God because God did both of these in Genesis 1.

A third aspect of being made "in the image of God" involves man's responsibility to manage the created world: "God said to them, 'Be fruitful and multiply, and fill the earth, and subdue it; and rule over the fish of the sea and over the birds of the sky and over every living thing that moves on the earth'" (Genesis 1:28). He merely told the animals to "Be fruitful and multiply, and fill the waters in the seas, and let birds multiply on the earth" (Genesis 1:22). Notice that we put apes in the zoo—not the other way around!

Fourth, the Bible says that humans were given a *position* in God's creation that is right below the angels and far above the animals:

> When I consider Your heavens, the work of Your fingers, the moon and the stars, which You have ordained; What is man that You take thought of him, and the son of man that You care for him? Yet You have made him a little lower than God, and You crown him with glory and majesty! You make him to rule over the works of Your hands; You have put all things under his feet, all sheep and oxen, and also the beasts of the field, the birds of the heavens and the fish of the sea, whatever passes through the paths of the seas (Psalm 8:3–8, NASB).

This passage shows that we are just below the angels and above the animal kingdom. Taking this a step further, the Bible even says that we *will judge the world and even the angels*: "Or do you not know that the saints will judge the world? If the world is judged by you, are you not competent to constitute the smallest law courts? Do you not know that we will judge angels? How much more matters of this life?" (1 Corinthians 6:2–3). This means that our "pecking order" in all of God's creation is very high—way above all of the created animals!

Fifth, and most importantly, God breathed His Spirit into humans: "Then the Lord God formed man of dust from the ground, and breathed into his nostrils the breath of life; and man became a living being" (Genesis 2:7). God breathed the breath of life into humans directly, rather than indirectly, as *imparted* to the animals. This breath represents the everlasting spirit that each of us has. The Bible clearly teaches that we will inherit everlasting bodies that never decay, and that each person will either live in Heaven or exist in Hell based on what he or she believes about Jesus (John 3:16–20). Indeed, humans are the most special of all God's creations: "All flesh is not the same flesh, but there is one flesh of men, and another flesh of beasts, and another flesh of birds, and another of fish" (1 Corinthians 15:39).

We can contrast these five aspects with the theory of human evolution. Jesus himself confirmed that humans were in fact created *at the beginning of creation* (Mark 10:6) and Paul confirms that Adam was created "as the first man" (1 Corinthians 15:45). Man did not evolve through a slow, molecules-to-man process according to God's Word.

Because humans are made in God's image, were the last beings God created, and were put in charge over His creation, it makes sense that humans today (and in the beginning) would be designed much differently than all other living things. For example, it makes sense that we live longer than most and that we are smarter, even being able to plan ahead. Truly, we were designed to take care of His creation. We are the only beings equipped for this role, even if we sometimes don't make the right choices.

TWELVE HUMAN DESIGN FEATURES THAT APES DON'T SHARE

Humans and apes differ in many clear ways, especially in their capabilities. A few of the capabilities and capacities that humans have that apes do not include advanced speech, mathematics, musicianship, worship, prayer, holding ceremonies, creativity, and love. These are just some of the things that humans can do that set us apart from every single animal on earth today. While each of these differences is important, humans reflect the *image of God* in another unique way: we can *create!* We cannot create things out of nothing like God, but we can creatively craft art and inventions with created materials. The ability to close our eyes and imagine new things, design, plan, and implement them is a uniquely human ability. Not only can we create and plan practical things like cars and planes, we create colorful and complex artwork! When is the last time you saw an ape engineer a bridge that is both strong and beautiful at the same time?

If humans were the result of evolution (rather than intentionally designed by God), shouldn't our physical functions and features center around survival—like hunting, gathering, fighting, escaping, etc? Clearly, humans are designed for much more than survival! We express creativity, joy, sorrow, and emotion, and we are capable of art, building, designing, writing, music, games, singing, and significance—not merely survival. Because we were created by God to rule and govern the animal kingdom, we would expect to find several mental and physical differences that God would bestow on us to help us with this responsibility. Now we will take a tour through twelve major design differences between humans and apes. As you're reading this section, remember that most evolutionists claim that human and chimp DNA is 99% similar[4]—an estimate which has subsequently been reduced to only 88%.[5] If the 99% estimate is true, and if DNA specifies traits, then what accounts for so many really significant differences between human and chimp traits?

DESIGN DIFFERENCE #1: THE "SOUL"

Physical differences between apes and humans are obvious, but non-physical differences may be even more significant. When you sat down at the zoo and watched primate behavior, you probably saw enough to understand that they behave like *animals*. That's why they are kept behind bars. People live and act on a whole different plane because we have *mind, heart, will*, and *conscience*. These four non-physical qualities combine to make humans incredibly unique.

By "mind" we mean self-awareness, mental states, beliefs, intentions, desires, and knowledge. These attributes work together in a way that enables human relationships to have a high degree of sophistication. They coordinate to make advanced predictions, and they understand and control our environment with a sense of time and planning. While some primates—chimps and orangutans—have a lesser version of some of these attributes, they are not even close to the equal of humans. By "heart," we mean emotions like appreciation, gratitude, and joy. Evidence exists that certain apes do show emotions, including expressing joy, anger, jealousy, compassion, despair, and affection,[6] but no ape has ever cried tears of joy. Even a novice observer can tell apes and humans have different hearts. By "will," we mean the power to choose, to show discipline and temperance, and to create and persevere. By "conscience," we mean an awareness of morality, or an overall sense of right and wrong. Our conscience lets us know when we fail to abide by either governmental laws or God's laws. Primates know nothing of laws. They live only by instincts.

DESIGN DIFFERENCE #2: THE SKULL

Apes and humans have vastly different skulls, and this is by design. For starters, the cranial capacity (the part of the skull that holds the brain) is much larger in humans because our brains are more than three times larger than the brains of chimps (for example, and also larger than every other primate—see Figure 7). The slope of our skulls is also much different than apes, as seen in Figure 2. In other words, humans have a more forward-oriented face in relation to the base of the skull. When an ape stands with its back straight up-and-down, it has to pull its head down toward its chest in order to face forward.

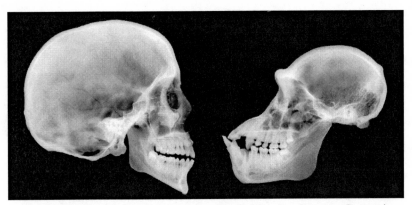

Figure 2. Chimp Skull v. Human Skull (Credit: Science Source)

Also, the face and jaws of humans lie underneath the brain case rather than protruding out like the ape. Some of the differences in the jaws, eye sockets, and skull shape between humans and apes are obvious yet subtle. One of the hidden (but significant) design differences between the skulls of both living and extinct apes and the skulls of humans is the location of the *foramen magnum*. This is the hole in the bottom of the skull where the spine enters. Because humans walk upright, the foramen magnum is located at the center of the bottom of the skull. In apes, the foramen magnum is located towards the back of the head. This way they can see straight ahead when walking on all fours. The figure below shows this distinction.

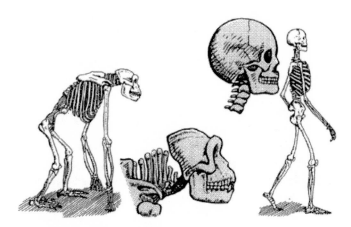

Figure 3. Human and Chimp Walking Angle Comparison (Credit: Evolution Facts, Inc. *Evolution Encyclopedia* Volume 2, Chapter 18 Ancient Man[7])

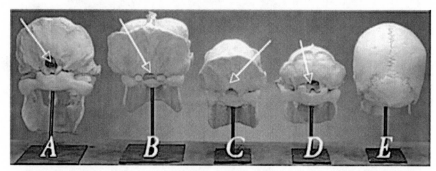

Figure 4. Foramen magnum (A: orangutan, B: male gorilla, C: female gorilla, D: chimp, E: human) (Credit: Answers in Genesis, *Image of God or Planet of the Apes*, 2006).

Figure 5. Foramen magnum, underside (A: orangutan, B: male gorilla, C: female gorilla, D: chimp, E: human) (Credit: Answers in Genesis, *Image of God or Planet of the Apes*, 2006).

There are several other major differences between ape and human skulls. Some of these include:

- The muscles that attach to the bottom of our skull that help us position, move, and stabilize the head.
- The **mastoid process**, which is a projection of our skulls behind the ear which provides an attachment for certain muscles of the neck.
- The **pre-maxilla,** which is a bone plate that bears the incisor teeth.
- **Facial prognathism,** or the angle of the face.
- Major differences in the **chin.**
- The **hyoid bone,** which is a special U-shaped bone that humans have just above the larynx.[8]

DESIGN DIFFERENCE #3: THE BRAIN

The difference between human and ape brains is incredible. In fact, the differences between our brains and virtually every ape-like creature alive today is probably one of the most significant proofs of Biblical Creation. Our brains are designed to lead, plan, control, make predictions and manage our lives, behaviors, choices, and environments—including those in the world around us—just as the Bible predicted. God gave us this superior brain power so we would be equipped to "rule over the fish of the sea and over the birds of the sky and over the cattle and over all the earth, and over every creeping thing that creeps on the earth" (Genesis 1:26 NASB).

Brains are the control centers for every involuntary action (like regulating our heart, breathing, perspiration) and voluntary action of every living organism. Humans have superior brains that are better equipped than any living organism when it comes to making carefully calculated actions based on predictions that are developed by our prior learning. Stop and think about it for a minute—just how much knowledge, planning, and prediction is necessary to send people to the moon? We have to have a *command* of knowledge and science in several areas to accomplish this feat. There's not a creature on the planet that even comes close to our abilities in this area.

To start exploring the differences between our brains and apes, we'll start with the most basic comparison: *size.* Figure 6 shows a basic comparison between human and chimp brains.

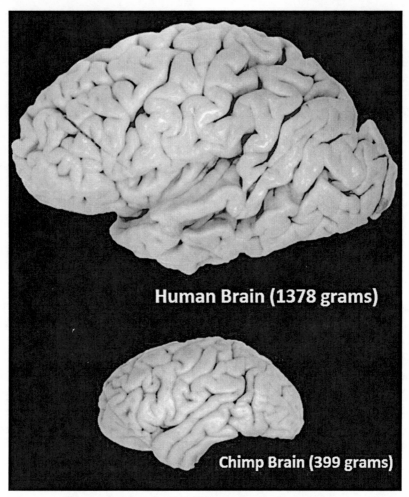

Human Brain (1378 grams)

Chimp Brain (399 grams)

Figure 6. Human and Chimp Brain Comparison
(Credit: Wikipedia Commons, brain weights added)

Figure 6 shows that human brains are about 3.5 times larger than modern chimp brains (399 grams v. 1378 grams on average).[9] Interestingly, the brains of human newborns typically range between 350 and 400 grams at birth—about the same size as a 100-pound, full-grown chimp brain! In fact, a chimp's tongue even weighs more than its brain![10] Many evolutionists consider chimps to be the "closest cousin" to humans, yet our brains are significantly larger and different.

Taking this a step further, Figure 7 shows 22 of the largest primates alive today (those weighing over 10 pounds), ranked according to their brain size.[11]

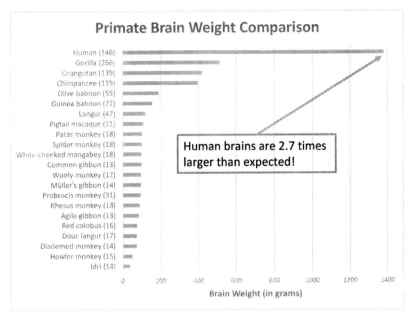

Figure 7. Primate Brain Weight Comparison

Figure 7 shows that the brain of a human weighs nearly three times more than the brain of a gorillas (1378 grams vs. 513 grams), while most humans weigh just about one-half of Gorillas (146 pounds vs. 266 pounds)! That's an enormous difference!

Next, let's take a look at this incredible difference in another way, using mathematical regression. Regression is a method for evaluating the correlation between two variables (in this case, the correlation between body weight and brain weight). Regression studies that result in correlations that exceed $r = .90$ demonstrate a *nearly perfect* association between two variables. This is exactly what occurs in the animal kingdom when studying the relationship between body weight and brain weight.

Dr. Thomas Schoenemann conducted an extensive study that evaluated the correlation between body weight and brain weight for over three hundred species and reported: "This relationship is quite strong across the vast range of mammalian brain and body sizes (the correlation is typically greater than $r = 0.95$)."[12] When the data in Figure 8 is used in a regression analysis to determine the correlation between the body and brain weights of the twenty-one primates listed (excluding humans), a correlation of this same strength ($r = .95$) emerges. However, including humans in the regression lowers the correlation to a value of only $r = .70$. It seems that the connection between body and brain weight is nearly perfect until humans are erroneously included in the primate group. At this point, the strong connection falls apart.

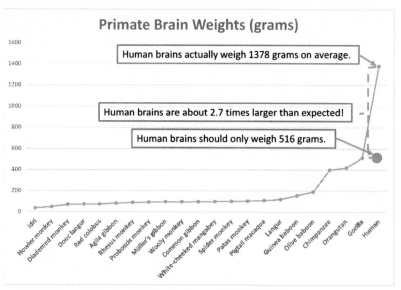

Figure 8. Primate v. Human Brain Weight. If humans are to be considered as one of the primates, our brains should weigh only 516 grams. In actuality, human brains on average weigh 1378 grams.

Based on the average body weight of 146 pounds, the predicted brain size of humans is 516 grams. This prediction represents a "miss" of about 862 grams (1378 actual v. 516 predicted). This is so far outside of the other primates that it is well beyond chance (the odds are 1 in 53,170!). Humans are far outside of the "animal kingdom" when it comes to brain size!

Why is it that only humans don't fit the model? Clearly the statistics show that all primates fit very well within their own category—the animal kingdom. Just like the Bible predicted, humans were made intelligent at the very beginning in order to fulfill our purpose of taking dominion over the earth.

In addition to the notable brain size difference, chimps and humans also show several significant **design differences**.

1. **DNA Methylation.** Scientists have discovered that human brain chemistry is very different than that of chimps. Human and chimp DNA methylation patterns in brain tissue (called "methylomes") are different.[13] DNA methylation is a biochemical process that helps

determine which genes will be more or less active. It occurs during development from an embryo through adulthood. As Institute for Creation Research Science Writer Brian Thomas points out, "If humans and chimps are close relatives, then they should have similar DNA methylation patterns in the areas of chromosomes that they have in common such as similar gene sequences. However, this team found major differences."[14] The "team" was made of evolutionary geneticists who, as usual, ran into evidence that they had to explain away, rather than the evidence they expected.

2. **The Temporal Cortex.** This part of our brain helps to process input from our senses and convert it into language. Which one has a largest temporal cortex? If you guessed "human," you are correct. Our temporal cortex would be larger than that of a chimp even if a chimp's heads could grow as large as a human's head![15] This shows that we were created from the very beginning to acquire, use and master language. We can even create new languages if we want to! The English language contains over a million words, and most people can learn and use most of them through a lifetime. But the vocal communications of chimps and all other primates are limited to noises, sounds, and grunts, amounting to (at most) tiny vocabularies.

3. **The Cerebellum.** The human brain cerebellum coordinates muscle motions involved in complex body movements, posture, and balance. We outdo chimps in this area of brain anatomy as well—humans are about three times larger than chimps.[16] Research has also shown that the human cerebellums are larger than expected, even if apes could grow to the size of a human.[17] Our superior cerebellum coordinates fine-tuned and delicate movements that no primate can perform. For example, only people can write a book, an essay, or even a sentence, using their fingers holding a pen.

4. **The Neocortex.** Distinguished Research Professor Jim Rilling (Emory University) has studied several human and chimp brains using Functional Magnetic Resonance Imaging (fMRI). These "deep scan" techniques measure brain activity by detecting changes associated with blood flow. Regarding the difference between human and chimp brains, he stated: "The human brain is not just an enlarged non-human primate brain, it is a different brain; one dominated by cerebral cortex." He notes that the human neocortex is disproportionately large compared to the rest of the brain.[18] In fact, there is only a 30:1 ratio of neocortical gray matter to the size of the medulla in the brainstem of chimpanzees, while that ratio is 60:1 in humans.[19]

5. **Spindle cells.** Neuroscientist John Allman and his team of scientists from Caltech[20] came out with breaking research results in 1999 that revealed a new kind of nerve cell in our brains called *spindle*

cells. Spindle cells were first found in the anterior cingulate cortex, a part of the brain that forms a "collar" around the corpus callosum (a part of the brain that connects the right and left hemispheres of the brain). The function and location of these spindle cells in this part of the brain is especially important because the work to *integrate* the parts of our brain that control the "automatic" functions of our bodies (like heart rate and nervous system) with "higher" brain functions such as attention and decision-making. This makes us superior to animals when it comes to making decisions "especially in the fact of conflicting information—anticipation of rewards, vocalization of language, and empathy."[21] Allman's team found that humans have almost twice as many of these specialized cells than chimps, and this helps set humans apart from chimps in many ways. For example, the special placement of these "long range" transmitters in the medial prefrontal cortex part of our brains helps enable us to "pull out memories from past experiences and use them to plot 'next moves.'" In particular, it becomes active when you present people with moral dilemmas in which their decision will directly affect the lives of others."[22] Doesn't this sound like a design feature that reflects the fact that we were "made in the image of God" as the Bible states?

6. **The Insula.** Among other things, this part of the brain is involved in taking information from our skin, internal organs, and cardiovascular system and converting them into subjective feelings such as empathy: "The anterior insula, together with the anterior cingulate is also involved in feelings of empathy toward others because both are particularly responsive to cries of pain and the sign of others in anguish or pain."[23] When counting spindle cells in this part of the brain, Allman's team (see above) found 82,855 in humans and only 1,808 in chimps. Thus, for every spindle cell found in the chimp's insula, humans have forty-six! This 46-to-1 ratio is reason to believe that humans are designed very differently than chimps. We are expressive, sensitive, empathetic, intuitive beings—not animals. Allman believes that the unique features in the human brain that pertain to the anterior cingulate, insula, and prefrontal cortex form a "neurological substrate for moral intuition... that can guide us highly complex, highly uncertain, and rapidly-changing social situations..."[24] Other brain researchers have confirmed the uniquely human features in the insula. For example, Christian Keysers has claimed that "complex emotions like guilt, shame, pride, embarrassment, disgust, and lust are based on a uniquely human mirror neuron system found *inside the insula*" (emphasis added).[25]

But what about the main function of the brain—smarts? Making effective choices? Figuring out mental challenges? In this area (intelligence), we beat chimps hands-down. Our brains *triple* in weight during childhood, enabling us to quickly grow in knowledge and increase physical skills. Humans have incredible memories, too, which enable us to have dominion over the animal kingdom. For example, the world record for memorizing numbers goes to Hiroyuki Goto of Tokyo, Japan, who was able to recite Pi from memory to 42,195 decimal places. Our superior intelligence allows humans to have the ability to have dominion over the animal kingdom.

Yes, apes are smart—but only smart enough to live and survive in their native environment. They show no evidence whatsoever of evolving or taking over the world, despite Hollywood movies that portray these fictional possibilities. Some researchers have famously taught gorillas to use sign language and picture symbols to communicate at a basic level, but their attempts cannot surpass these limited abilities. While humans can convey meaning, ideas, and mathematical formulas using language, and can learn many different languages, communication between chimps in the wild is limited to certain facial expressions, gestures, and a variety of screams, hoots, and roars.

Chimps can use sticks to "fish" for termites. They can crack nuts open using stones against hard surfaces. They can use leafy sponges to drink water. They're even smart enough to use a stick when prodding snakes to be sure they are dead. They will use leaves to wipe themselves. Meanwhile, humans build libraries and fill them with books, go to the moon and back, invent advancements in medicine and technology, consider our origins, and worship our Creator by serving others.

We've been using intelligence tests for over 100 years to stack up our smarts against others. Popular intelligence tests result with an Intelligence Quotient (or "IQ") score indicating how smart we are. These test scores can show our mental abilities in specific areas, such as verbal comprehension, reasoning, vocabulary, information processing, spatial rotations (i.e., working with blocks and puzzles), numerical computations, and processing speed. The average IQ score for humans is 100, and about 52% of humans score between 90 and 110. Only about 2% of people score below 70 and only 2% score over 130. But if an IQ test was given to the world's smartest chimp, it wouldn't even be able to register any score on any of the tests! Chimps lack even the ability to understand the directions.

Some scientists have attempted to show that chimps can beat humans on some basic mental routines such as memorizing numbers. One such study was conducted by the Primate Research Institute of Kyoto University in central Japan.[26] The study alleged to show that chimps could beat humans in tests that measured "working memory" using "numbers" (they weren't actually numbers to the chimps, but rather just shapes because the chimps were not tested on the ordinal values).

The media ran the story around the world, using it as evidence of the close human-chimp relationship and evolutionary connection (a simple internet search reveals numerous hits). News stories, press releases, and interviews were everywhere. Then, a few years after this "amazing" results, the study was finally analyzed by other scientists, and quickly debunked. For example, in 2009, Siberberg and Kearns[27] discovered that the original study was critically flawed because the chimps were given several practice sessions while none of the human participants had any practice sessions. This explained the difference in memory performance. Part of their study showed that humans could perform at the same level as the leading "memory chimp" when given even moderate levels of practice.

By this time, however, the "chimps beat humans at memory" study had already become so popular that most people had already accepted the idea that chimp working memory was superior. The myth even continued in the popular press, with several online news sources running specials on the "super memory" chimps in 2012, and a BBC documentary titled *Super Smart Animals* was released which highlighted the leading chimp's memory abilities. The academic field, however, continued to show the methodological flaws in the original study, including the fact that the chimps in the study had been trained on skills related to the test for *nearly a decade* before the study was conducted.[28]

Even if chimps exceed human abilities in certain areas, this does not in any way challenge the Creationist position. Animals have been shown to exceed human abilities in many ways! Chimps do have a great memory for recalling shapes and patterns quickly. Sea lions and elephants can remember faces from decades ago. Certain dogs can even use their exceptional sense of smell for detecting early signs of colon cancer. Humming birds can travel hundreds of miles using a sugar cube worth of energy. There is no doubt, however, that the Creator of the Universe has endowed humans with the intellectual capacity that exceeds all life on earth, and that was part of His design for humans to take dominion over the Earth: "The heaven, even the heavens, are the Lord's; But the earth He has given to the children of men" (Psalms 115:16).

DESIGN DIFFERENCE #4: THE EARS

Even our ears are different than those of gorillas, gibbons, monkeys, and chimps. Deep inside of our inner ears is a set of three tubes called "semicircular canals" specially designed to aid balance. Semicircular canals in humans are oriented in our skulls specifically for walking upright while facing forward. The size, shape, and 3-D orientation of semicircular canals matches each animal's skull orientation, keeping them balanced. Like many important body parts, disabled semicircular canals illustrate their function. When an animal suffers an untreated inner ear infection that destroys the semicircular canals in only one ear but then recovers, it spends the rest of its days with its head tilted sideways. Animals that have fast, jerky movements have larger semicircular canals relative to their body size than those that move more cautiously.[29]

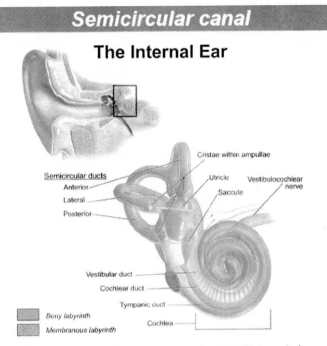

Figure 9. Semicircular Canal (Credit: Wikipedia)

Scientists have researched human and chimp semicircular canals to see how different they are in shape and design. These studies reveal both *visibly* and *statistically* obvious differences, with one study showing that humans and two different primates differ with a 99.99% certainty![30] Truly, humans and primates were made different from the start.

What difference does this make? Well, think about this: If you were to go in for surgery and have your semicircular canals replaced with those from a chimp, at the very least, you'd be confused and disoriented. You certainly wouldn't be able to run with as much ease as you have now! This is because the semicircular canals are critical for maintaining balance and, in humans, two of the three help stabilize your head when running.[31]

DESIGN DIFFERENCE #5: THE FACE, MOUTH, AND SPEECH

Take a minute to stand up and look at something straight ahead that is about level with your eyes. Then look straight down. You'll notice that you can look at the ground right at your feet. If you were a chimp, you wouldn't be looking at the ground—you'd be looking down at your own face! This is because chimps have a sloped face compared to humans, and this is by design because they walk on all fours.

Another obvious facial difference between humans and chimps is the eyes. Unlike *all* of the more than two hundred species of primates and other animals, humans have whites around the eyes. This part of our eyes, called the sclera, combined with the colored iris, and the black pupil, distinguish human eyes from the eyes of all other living beings.

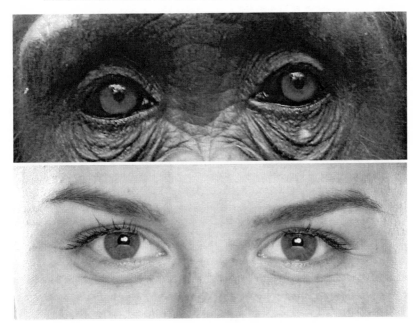

Figure 10. Human v. Primate Eyes (Credit: Shuttertock)

One might think, "There's no big deal to having 'whites' in our eyes!" Think again! The whites of our eyes allows us to know who is looking at us even in a room filled with people. If you look at a chimp in the zoo, it is difficult to tell if they are looking at you or at someone else! We were designed to perceive eyeball movements, whereas animals merely notice whole face movements. Who hasn't seen someone roll their eyes in disgust, or widen their eyes in surprise? Our scleras open a whole new dimension of communication that no animal has.

Scientists have acknowledged that this distinction has enabled humans to work more cooperatively in teams compared to primates.[32] In fact, experiments reveal that human infants at the one year of age (before they can talk or understand spoken language) tend to follow the direction of another person's eyes instead of following their heads. For example, one study showed that when a mother is looking directly at her baby, and then moves her eyes upward toward the ceiling, the baby also looked up toward the ceiling. But when the mom closes her eyes and points her head to the ceiling as if she was looking upward, the baby does not typically follow the motion of her head. This study also showed that chimpanzees, bonobos, and gorillas showed *precisely the opposite* pattern of gaze.[33]

DESIGN DIFFERENCE #6: THE FACIAL MUSCLES

Humans have fifty facial muscles and the unique ability to make over ten thousand different facial expressions.[34] Chimps only have twenty-three facial muscles—only one-half that of humans![35] This profound difference, combined with the fact that humans have the ability to speak, makes humans superior communicators. The fact that we can speak also shows our unique design in the image of God, who spoke all of creation into existence, who speaks to us through His Word, and who listens when we speak to Him in prayer.

Chimps would require a complete overhaul in order to speak like we do. Not only would you need to add several more facial muscles to control the lips and cheeks for making certain sounds, you'd also have to redesign the vocal chords to connect with a newly formed hyoid bone, construct and integrate a deeper, and more resonant pharynx, engineer finer lip controls for releasing pressure, and craft specific changes to the tongue. Of course, you'd also have to rewire those parts of the chimp's brain that conduct and coordinate speech and other expressions during communication. The parts of the human brain responsible for handling speech, called the Brodmann areas 44 and 45, are over six times larger in humans compared to chimps.[36]

These differences were given to humans because we were made in God's image with the responsibility to be caretakers over the earth and to relate to one another and to our Savior. We have every communication feature we would expect to see if we were designed by God.

DESIGN DIFFERENCE #7: THE SPINE

Humans have a spine with two gradual "S"-shaped curves. No other creature uses this unique double curve. It brings the human head and torso into a straight line above our feet so we can walk and run long distances. Apes most naturally walk on all fours, so they have a back that supports their arms and hands for walking.

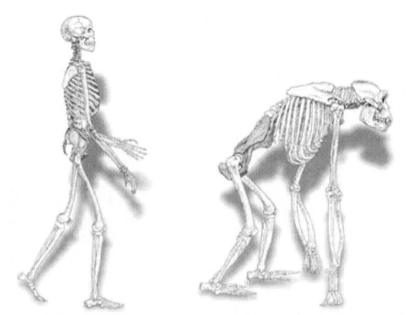

Figure 11. Human and Chimp Stride Comparison. This image shows the S-shape in the human spine and the bow-shaped chimp spine. Truly we are very different than chimps by design![37]

Apes, both those in the fossil record and those alive today, are each different from humans. Chimps, for example, have one more thoracic vertebra, one less lumbar vertebra, and one less caudal vertebra than do humans. Normally both have seven cervical vertebrae and the combined thoracic, lumbar, and sacral regions consist of twenty-two vertebrae. Chimps lack the extreme curves of the human column, and the angle between the lumbar and sacral region is more acute in a chimp.[38]

Why do these differences matter? The spine anchors and connects all the other body regions. Clearly, the human spine was crafted to support and keep in proper balance the uniquely human legs, arms and head. The same could be said for each separately created species of ape, whether living or extinct. Evolution demands that an ape-like ancestor morphed into mankind. If so, which parts morphed first? What if it was the spine? Connecting ape-like legs, arms and a head to a human spine would create a creature that could not walk like a human or like an ape. It would die. In this way, we know that God created each separate kind.

DESIGN DIFFERENCE #8: THE WRIST AND HAND

Have you ever tried walking on your hands? It's not easy. One of the main difficulties is that our wrists don't have locking bone and joint structures. The wrists of both modern apes[39] and extinct australopithecine apes lock in place. This mechanism supports their body weight during knuckle walking.[40] Take a look at their wrist designs below.

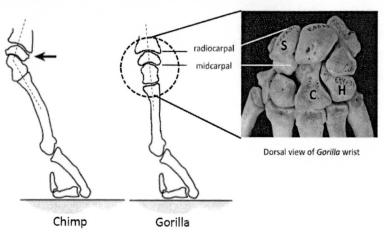

Dorsal view of *Gorilla* wrist

Figure 12. Chimp and Gorilla Wrists[41]

The top arrow in Figure 12 shows *ridges* on the radius and scaphoid bones and the bottom arrow shows the wrist and finger bones in chimps. These ridges act as stoppers to keep these bones stable and aligned beneath a chimpanzee's body weight when walking on their hands. The right side of the image shows similar features on the gorilla's hand but not on the gorilla's wrist. These features are not present in humans.

Chimp and human hands are also very different. Our hands are much more mobile and flexible than chimp hands. We can completely rotate our hands and extend and flex our hands at the wrist. Most primates (especially those that walk on their knuckles), are not as flexible with their hand movements. In part, this is because their wrist bones prevent their hands from bending, or extending, while putting pressure on their knuckles. In addition, the hand bones and finger bones of most ape-like creatures are curved because they are designed to swing from tree branches. [42] This anatomy keeps them from being able to manipulate tools like the uniquely human hand can.[43] Only humans can chop wood with an axe and perform delicate eye surgery with the same hands.

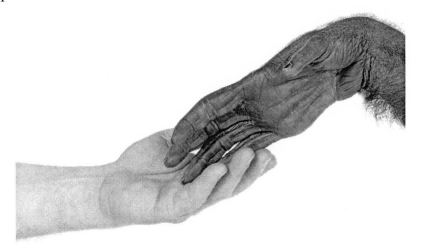

Figure 13. Human and Chimp Hands (Credit: Shutterstock)

Even our thumbs are different than those of apes. Because our thumbs are longer and stronger than apes, our grip is stronger and much more precise than apes. This allows us to make and use finer and more precise motions and tools than apes.[44] In addition, humans have "opposable thumbs," meaning that we can pinch items between our thumbs and each fingertip. Those primates that do have a version of opposable thumbs still lack the uniquely human arm muscles. Finally, primates—which we understand to exclude people—do not have uniquely human nerve controls that guide slow and delicate motions that no animal can perform.[45] Truly we are designed to work, serve, worship, and have dominion over the earth!

DESIGN DIFFERENCE #9: THE HIPS/PELVIS

The way that humans walk and run depend on the movement of our hips, legs, ribs, legs, and arms. The shape, size, and configuration of our bones, muscles, tendons, and ligaments also play a huge role in our unique ability to walk and run upright (compared to chimps, who typically walk on all fours).

The hips are at the top of this walking system. Our wing-like pelvic bones (the iliac blades) are *curved* so a certain muscle that's attached to it—the gluteus medius—can balance our legs so we can both walk and stand upright. Chimps and gorillas have a much different design, as described by Dr. Elizabeth Mitchell: "Chimpanzees and gorillas, unlike humans, have *flat* iliac blades; therefore, their gluteus medius muscles are oriented differently. While walking on four legs, their backward-oriented muscles keep them from falling forward on their faces. But without curved iliac blades, the muscles cannot stabilize the planted leg during bipedal stepping." [46]

In addition, chimps have a *longer* ilium (the uppermost and largest bone of the pelvis) than humans. Their hips orient their legs so that their knees point outward. This gives them better range of motion for climbing, but if they tried to walk upright like humans, they would waddle almost painfully. In contrast, the human hips anchor their legs with knees pointing straight ahead. This structure helps us walk and run all day long— assuming we're in good shape. Figure 14 shows how chimps are uniquely designed for walking on all fours, and Figure 15 shows the unique design of human hips.

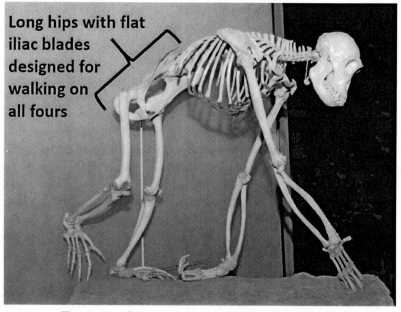

Figure 14. Chimp Hips (Credit: Science Source)

Human hips have shorter, curved iliac blades so the gluteus medius can attach in a way for upright walking

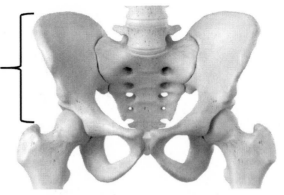

Figure 15. Human Hips (Credit: Shutterstock)

DESIGN DIFFERENCE #10: THE LEGS

Only humans stand and walk entirely on two feet. Kangaroos can stand on two feet, but they hop rather than walk and their forepaws are too small for any mobility function. Chimps can walk on two feet, but they do not do that very efficiently nor for very long. They are clearly designed for quadrupedalism, or the ability to move around on four feet. Why is this an important distinction? There are several reasons. Remember, God created humans to have dominion over the rest of the animal kingdom, and He gave us the functional means in our bodies to help with that responsibility. Here are some of the advantages of bipedalism over quadrupedalism:

- Walking upright gives us free hands to use tools, even while walking. This way, we can use a bow and arrow or text message while on our feet.
- Walking on two feet is more efficient than walking on all fours. By saving energy, we can walk all day without getting too tired.
- Taking care of babies. Chimp infants cling to their tree-dwelling mothers, but human babies often need both of their mommy's hands.
- Increased height for viewing across landscape.

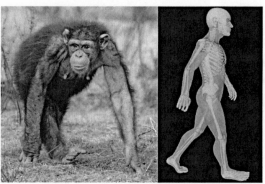

Figure 16. Ape and Human Walking Posture (Credit: Shutterstock)

DESIGN DIFFERENCE #11: THE FEET

As described by Dr. Elizabeth Mitchell, our feet are perfectly designed for walking and running:

> As the foot is planted, many bones lock to form rigid weight-bearing levers able to transfer our weight as we rock forward. The 26 perfectly-shaped foot bones with their ligaments, tendons, and arches keep the foot loose enough to absorb shock and adapt to uneven surfaces while remaining stable enough to support the human body's full weight. The curvature of our back also aids balance.[47]

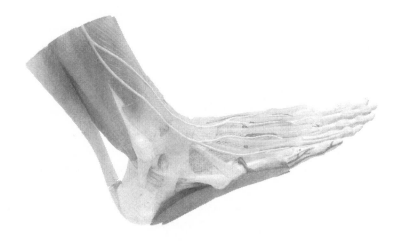

Figure 17. Human Feet (Credit: Shutterstock)

Apes, on the other hand, have *hands* for feet. The large toe of a chimp foot *opposes* its other toes so they can clamp onto branches and vines when climbing. The large toe of human feet is *aligned* with other digits for walking and running, and can't grasp much at all!

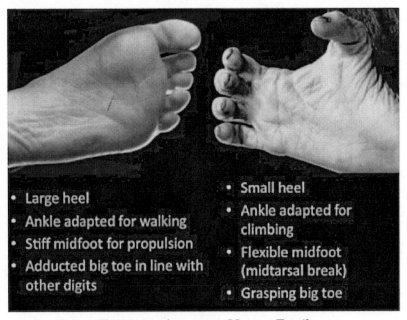

- Large heel
- Ankle adapted for walking
- Stiff midfoot for propulsion
- Adducted big toe in line with other digits

- Small heel
- Ankle adapted for climbing
- Flexible midfoot (midtarsal break)
- Grasping big toe

Figure 18. Ape versus Human Feet[48]

Ape feet have a small heel, a grasping big toe, and a flexible hinge right in the middle of each foot. Human feet line up with our big toe for "pushing off" when running. In addition, the ankle bones (tarsals) of human feet are larger and more rigid than the chimps. Of all primates, only bonobos are capable of walking confidently on two feet (which they do only about 25% of the time). Chimps are only able to walk confidently on two feet when they are in the water. All other primate species are usually inclined to use their "foot-hand" system of walking.

The human foot is also arched, which gives us balance and stability when walking. These arches literally put a spring in our steps. Apes have no arch, and walking flat-footed makes their feet more like hands—great for climbing, not for walking. When it comes to traveling long distances, humans have chimps beat by a factor of three to four times! While estimates vary, many humans can travel twenty to thirty miles in a day (and some much further), while chimps can only cover around six miles.[49] Traveling such distances is only possible because of our well-integrated spine, hips, legs, and even weight-balancing arms and forward facing heads—all of which enable us to walk upright.

DESIGN DIFFERENCE #12: THE SKIN

Why don't humans have fur? The reason is because our skin is entirely different than that of apes! The most obvious difference is that our skin is not completely covered with hair like apes. We also have more sweat glands, produce vitamin D, and have melanin that makes our different skin colors. Our fine touch sensors make human skin ideally suited for affection and touching.

Dr. Montagna works at the Department of Cutaneous Biology at the Oregon Regional Primate Research Center. He conducted an extensive research study on the skin differences between humans and apes, finding some intriguing details:

- All nonhuman primates (an evolutionary way of saying "primates") have a hairy coat, which can be thick or thin, short or long, woolly or shaggy, dense or sparse, and show varied colors.
- All primates have fingerprints on their friction surfaces; they are also present on the tail of some monkeys and on the knuckle pads of chimpanzees and gorillas.
- Their skin is thin all around and has little structure underneath.

- The friction surfaces of most nonhuman primate species have sweat glands. Sweat glands in chimps and gorillas do not respond to heat stimulation as they do in man.[50]

SOME OTHER IMPORTANT DESIGN DIFFERENCES BETWEEN HUMANS AND APES

We've covered some major design differences between humans and apes, but many more deserve our attention:

- On average, humans are about 38% taller, 80% heavier, live 50% longer, and have brains that are about 400% larger (1330 ccs compared to 330 ccs).[51]
- Chimps show aggression by showing their teeth; people smile showing their teeth to show warmth.
- Certain parrots have larger vocabularies and more elaborate language skills than any primate.
- When it comes to sexual reproduction and relationships, only humans experience jealousy or competition with "sheer joy" and excitement; chimps typically mate with multiple short-term partners.
- Humans design and use highly complex tools and multi-component systems; chimps only use basic tools, and *they do not even do that as cleverly as crows do!*
- Humans adapt their surroundings to themselves; chimps adapt themselves to their surroundings.
- Humans have directed and systematic ways for educating the next generation; education is mostly indirect and instinctual with chimps.

Even if human and chimpanzee DNA sequences are similar, which they are not, we have shown that they are much more different than commonly taught. Their DNA-coded information produces very different creatures!

CONCLUSION

If human evolution was true, then we should find millions, thousands, hundreds, or maybe even just a dozen "in-between" creatures alive today. But the score for these "transitional forms" today is *zero*. Not one creature lives today that can be branded half-ape and half-human. Instead, apes produce apes and humans produce humans, just like the Bible describes.

If "molecules-to-man" evolution were true, we would expect evidence of millions of in-between species—creatures that were still evolving along the ape-to-human progression. This was even a question that Charles Darwin, the 19th century promoter of evolution, asked about the historical fossil record. He wrote, "By evolution theory, innumerable transitional forms must have existed." He then asked, "Why do we not find them embedded in countless numbers in the crust of the earth?"[52] And, if human evolution were true, wouldn't we expect plenty of obvious "transitional creatures" between apes and humans in the crust of the earth? Darwin also asked, "Why is not every geological formation and every stratum full of such intermediate links? Geology assuredly does not reveal any such finely graduated organic chain." He followed this question by saying this was the "most obvious and serious objection which can be urged against the theory."[53] In other words, Darwin knew that his theory would be weakened if researchers in the future did not dig up millions of "in-between" creatures.

But what do we find today? Millions of ape-human half-breeds? No. In fact, we see over seven billion people on the planet who are all obviously human in every sense. Though some physical and many cultural differences display God's creativity, we are all the same kind—sons and daughters of Adam and Eve. We are all inter-fertile. And what do we find in the crust of the earth after digging up billions of fossils for over 150 years since Darwin posed his "big questions" above? We find a handful of fossil creatures that better fit either the apes or human categories than they do the evolutionary category of ape-human transition.

It seems that Darwin wanted a *clear line of evidence* showing "half-way-in-between" ape-human creatures. These transitions should be alive today, and they should have left millions of their bones in earth's sediments. Yet we do not find either form of evidence. What we see instead is what God said: mankind is made in the image and likeness of God, able to think, plan, worship, pray, and create. We also see variations *between* people groups as the Bible mentioned both in the Old (Genesis 9:18–

19) and New Testaments. For example, consider Acts 17:26–27, which records Paul's gospel presentation to pagans. It says, "And He has made from one blood every nation of men to dwell on all the face of the earth, and has determined their pre-appointed times and the boundaries of their dwellings, so that they should seek the Lord, in the hope that they might grope for Him and find Him, though He is not far from each one of us."

God made humans in His likeness, breathed His breath of life in them, and gave them charge over all of His Creation—just as we still see today. We see today just what the Bible specified ten times in the very first chapter: that all living creatures would reproduce and fill the earth each after their own kind.

"Know that the Lord is God. It is he who made us, and we are his; we are his people, the sheep of his pasture."
– Psalm 100:3

CHAPTER 2

A Brief History of Human Evolution in Textbooks

Daniel A. Biddle, Ph.D.

EVOLUTION'S BEGINNINGS

The standard line-up of the four ape-to-human icons that public school textbooks most often feature today is quite different from the evolutionary story of apes progressing to humans described in the past. The standard lineup presented in textbooks changes at least every couple of decades. If ideas of human evolution are false, we would expect them to shift frequently, just as history has proven happens. To demonstrate this, let's journey through time and review the once-best, but now discarded, evolutionary ideas that perhaps your grandparents' textbooks promoted.

In **1829**, Neanderthal skulls were first discovered in Belgium, and dozens have been found since. Originally classified as "pre-humans" or "sub-humans," they are now believed to be human in every practical sense. These ancient humans had unique features, but none that lie outside the range of modern men and women. Nevertheless, Neanderthal's peculiarities were too tempting for those anxious to find a missing link. They thought they found it. However, recent discoveries prove that Neanderthals were fully human—descendants of Adam just like us. They buried their dead, made instruments, practiced burial rituals, and made and used advanced tools. They have even been found buried alongside modern-looking humans.[54] Their skulls were close to 200 cc greater than that of present-day humans—hardly an intermediate form between humans and apes!

Neanderthal expert Erik Trinkhaus admitted, "Detailed comparisons of Neanderthal skeletal remains with those of modern humans have shown that there is nothing in Neanderthal anatomy that conclusively indicates locomotor, manipulative, intellectual, or linguistic abilities inferior to those of modern humans."[55] If textbook writers of yesteryear had waited until evolutionists examined Neanderthal fossils enough to see that they were fully human, they would not have been able to illustrate human evolution very well.

Figures 19 and 20 show this changing position on Neanderthals—from pre-human "brute" to human.

Figure 19. Previous Idea of Neanderthal Man (Credit: This reconstruction of the La Chapelle-aux-Saints Neanderthal skeleton—discovered in France in 1908—was published in *L'Illustration and in the Illustrated London News* in 1909).

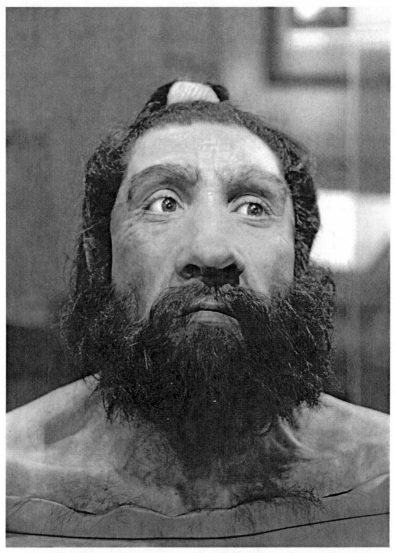

Figure 20. Current Idea of Neanderthal Man. Once considered an ape-like caveman, Neanderthal remains have proven their identity as fully human. Give him a shave, haircut, and button-down shirt and this Neanderthal would blend right into a city crowd (Credit: Wikipedia).

In **1859,** Charles Darwin published the *Origin of Species by Means of Natural Selection.* This book did not broach the topic of how evolution might apply to humans. Darwin only stated that future research would reveal the origin of man: "light will be thrown on the origin of man and his history" (Chapter 14).

In **1863**, a famous promoter of evolution Thomas Henry Huxley, laid out his best case to show that humans evolved from apes in a book titled *Evidence as to Man's Place in Nature*.[56] In his book, Huxley concluded, "it is quite certain that the Ape which most nearly approaches man, in the totality of his organization, is either the chimpanzee or the gorilla." Huxley presented one of the earliest "March of Man" images used to suggest human evolution (Figure 21).

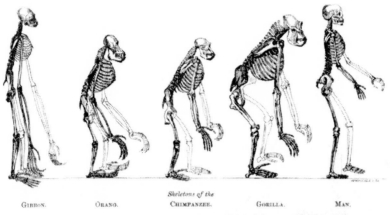

Skeletons of the

GIBBON.   ORANG.   CHIMPANZEE.   GORILLA.   MAN.

Photographically reduced from Diagrams of the natural size (except that of the Gibbon, which was twice as large as nature),
drawn by Mr. Waterhouse Hawkins from specimens in the Museum of the Royal College of Surgeons.

Figure 21. Huxley's Comparison of Ape and Human Skeletons (*Evidence as to Man's Place in Nature*, 1863). In contrast to Huxley's original caption, the "Man" skeleton is smaller in relation to chimp and orangutan. Also, these drawings depict awkward postures that make them look more similar than their natural postures would suggest (Credit: Wikipedia).

In **1871**, Darwin published *The Descent of Man*, in which he laid out his theory that humans are descended from ape-like creatures. Darwin supported his ideas from three main categories: similarities between humans and other primates, similarities in embryological development, and similarities in vestigial organs (which are parts of our bodies that are supposedly "leftover" from evolution). Darwin concludes that we are closely related to either gorillas or chimpanzees: "In each great region of the world the living mammals are closely related to the extinct species of the same region. It is, therefore, probable that Africa was formerly inhabited by extinct apes closely allied to the gorilla and chimpanzee; and as these two species are now man's nearest allies, it is somewhat more probable that our early progenitors lived on the African continent than elsewhere" (Darwin, *Decent of Man*, 1871).

Darwin's ideas bolstered the racist thoughts and ideas of the 19th and 20th centuries, and in some cases still today. Darwin's infamous book *Origin of the Species* was originally released in 1859 under the full title, *On the Origin of Species by Means of Natural Selection or the Preservation of Favoured Races in the Struggle for Life*. This title was shortened in 1872 (with the release of the sixth edition) to simply, *The Origin of Species*. Darwin's second book, *The Descent of Man*, included one chapter titled "The Races of Man." In this chapter, Darwin stated:

> At some future period not very distant as measured by centuries, the civilized races of man will almost certainly exterminate and replace the savage races throughout the world. At the same time, the anthropomorphous apes...will no doubt be exterminated. The break between man and his nearest Allies will then be wider, for it will intervene between man in a more civilized state, as we may hope, even than the Caucasian, and some ape as low as the baboon, instead of as now between the Negro or Australian and the gorilla.[57]

In chapter 7 he noted:

> Their mental characteristics are likewise very distinct; chiefly as it would appear in their emotional, but partly in their intellectual faculties. Everyone who has had the opportunity of comparison must have been struck with the contrast between the taciturn, even morose, aborigines of S. America and the light-hearted, talkative negroes.

Darwin's belief in racial superiority was obvious: If man evolved then so did the various races, and the "Caucasian" race evolved farther than others. The impact of these philosophies is enormous according to historians, who have traced Darwin's ideas to Hitler's death camps during World War II.[58]

In **1874**, Ernst Haeckel published *The Evolution of Man* which included a famous figure showing humans evolving from Amoeba to modern man through twenty-four stages.

THE MODERN THEORY OF THE DESCENT OF MAN.

Figure 22. Human Evolution Ideas in 1874
The figure shows humans evolving through twenty-four stages, from
Amoeba (1) to Worm (7) to Jawless Fish (lamprey) (10), to a Plesiosaur
(14), to Monkey (20), to Modern Human (24) (Credit: The modern
theory of the descent of man, by Ernst Haeckel, published in *The
Evolution of Man*, 1874).

While Biblical Creation continued as the predominant teaching
in public schools, evolutionary ideas began their first introductions into
school textbooks between 1888 and 1890.[59] Darwin published his last
work in 1882, the same year he died. Two complete Neanderthal skeletons
were found in 1886 in a cave in Belgium, giving naturalists fuel for more

evolutionary imaginings.

In **1891**, Ernst Haeckel updated his ideas about human evolution by publishing a new book titled *Anthropogeny*,[60] which included one of the earliest "trees" of human evolution. The trees change with the release of almost every new paper or research study, whether evolutionists use DNA sequences or body forms to guess at "relatedness." As you read this section, pay careful attention at how this "tree" changes.

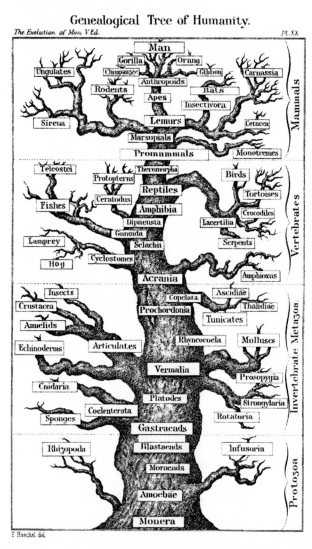

Figure 23. Ernst Haeckel's late 19th century (1891)[61] idea of which animal forms may have evolved into which over imagined eons.

JAVA MAN

Even bigger news came in **1891**, when Eugene Dubois enlisted the help of the colonial government, two engineers, and fifty convicts to manually tear through tons of earth on the Indonesian island of Java in an attempt to find "the missing link" between apes and humans.[62] In addition to numerous animal fossils, Dubois' team discovered a tooth, a skullcap, and a femur (thighbone) in East Java. While the femur was found a year later and about 50 feet from the skullcap, he assumed they were from the same creature. Dubois named the collection "Java Man" and gave it the scientific name *Pithecanthropus erectus*.

Immediately after he published his finds, the science community opposed them. When Java Man was presented before the Berlin Anthropological Society in January 1895, German Dr. W. Krause unhesitatingly declared that the tooth was a molar of an ape, the skull was from a gibbon, and the femur was human. Krause said, "The three could not belong to the same individual."[63] Despite reasonable objections, almost eighty books or articles had been published on Java Man within ten years of Dubois' find, explaining them as missing links for human evolution.

Decades of hype finally began to topple in 1939 when two experts, Ralph von Koenigswald and Franz Weidenreich, revealed that Java Man looked similar to a set of fossils found in 1923–1927 called "Peking Man," or *Sinanthropus pekinensis*. Both were actually normal human beings.[64] The final nail was hammered into the coffin of Java Man as a transitional form in 1944. Harvard University professor Ernst Mayr, the leading evolutionary biologist of the 20th century, classified both of these finds as human.[65]

Interestingly, Dubois found two definitely human skulls called the Wadjak skulls, which were discovered in strata at the same level as the "Java Man" fossils. Why did he keep them secret for thirty years? During that time, the international scientific community accepted Java Man as a real missing link.[66] Near the end of his life, however, Dubois publicly conceded that Java Man was extremely similar to (though he believed not identical with) a large gibbon. Dubois wrote, "*Pithecanthropus* was not a man, but a gigantic genus allied to the Gibbons,"[67] a statement over which both Creationists and Evolutionists are still quarreling.

What were the Java Man remains? They probably consisted of a human femur and extinct ape bones including a gibbon's skull remains. These simply show that some people and some apes were fossilized as distinct kinds. Java Man never really existed. Unfortunately, the next generation of public school textbooks did not admit this. Instead, they slyly replaced the "Java Man" story with a new hopeful evolutionary link.

Figure 24. Reconstruction of Java Man.[68] The white parts of the skull and the facial reconstruction was based only on the skullcap, which is the dark part on the top.

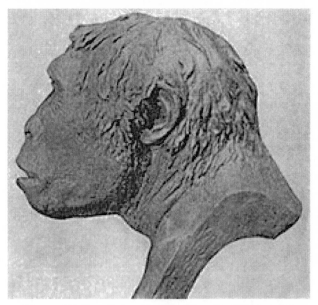

Figure 25. Java Man Profile (Credit: Wikipedia)

This Java Man profile has been prominently displayed in books and other media for decades.

Figure 26. Java Man Statue.[69] A statue of an imaginary reconstruction of a "Java Man" skull marks the land of its discovery, even though most evolutionists finally determined that it was no missing link at all.

WHAT WAS FISHY ABOUT THIS APE MAN?

Howard E. Wilson points out some interesting facts about Java Man—some that are not widely known.[70] Apparently, DuBois did not enjoy having people come view the actual Java Man fossils. He kept them under tight lock-and-key for thirty years. When others finally viewed them, the bones turned out to be vastly different than the copies displayed and analyzed around the world! The well-known journal *Science* published an article[71] that stated:

> There is a "skeleton in the closet" of man's evolutionary history, and Prof. E. DuBois… holds the key. The "closet" is said to be a good stout safe in Haarlem, Holland, and the skeleton is none other than that of *Pithecanthropus erectus*, the famous ape-man who [supposedly] lived in Java over a half million years ago. For thirty years scientists from all over Europe have besieged Dr. DuBois for permission to examine the remains, while eminent anthropologists have crossed the ocean for that purpose only to be turned away at the door.

After being largely hidden away for thirty years, Dr. Alex Hrdlicka of the Smithsonian Institute wrote, "None of the published illustrations or the casts now in various institutions is accurate. Especially is this true of the teeth and the thigh bone. The new brain cast is very close to human. The femur is without question human."[72] Now that's an amazing statement! "None of the published illustrations of the cast now in various institutions is accurate." How can Java Man be trusted as a "transition" between apes and man if none of the casts were accurate?

Some have argued that Java Man's brain was too small to be human. The cranial capacity of Java Man was estimated at 1,000 cc. This would be small, but fits within the range of modern humans. Apes never exceed 600 cc.[73] While 1,000 ccs is not a large cranial capacity, some people groups also have smaller brains (in the 900 to 1,000 cc range).[74] However, their diminished cranial capacity does not make them any less human nor any less intelligent.[75]

When the 20th century began, Biblical creation continued as the primary teaching on origins. Evolution teaching was scarcely taught in public schools. Oscar Richards more recently conducted a study of six of the most commonly used textbooks in the U.S. published between 1911 and 1919 (representing 75% of all U. S. schools). Using word counts, he estimated that only 1.68% of these textbooks was devoted to evolution.[76]

PILTDOWN MAN

"Piltdown Man" is a fraudulent composite of fossil human skull fragments plus a modern ape jaw with two teeth that Charles Dawson supposedly discovered in a gravel pit at Piltdown, east Sussex, England. History testifies, as summarized by Pat Shipman, that "the Piltdown fossils, whose discovery was first announced in 1912, fooled many of the greatest minds in paleoanthropology until 1953, when the remains were revealed as planted, altered—a forgery."[77]

The New York Times.

SUNDAY, DECEMBER 22, 1912.

DARWIN THEORY IS PROVED TRUE

English Scientists Say the Skull Found in Sussex Establishes Human Descent from Apes.

THOUGHT TO BE A WOMAN'S

Bones Illustrate a Stage of Evolution Which Has Only Been Imagined Before.

Figure 27. Piltdown Man Announced in the New York Times[78] (1912) Major media outlets have a long history of splashing headlines that support evolution, but burying news that refutes it. Piltdown Man was later proven a 100% fraud.

Consider the following deliberate (and desperate) measures some have used to promote belief in macro-evolution:

> Piltdown Common had been used as a mass grave during the great plagues of A.D. 1348–9. The skull bones were quite thick, a characteristic of more ancient fossils, and *the skull had been treated with potassium bichromate* by Dawson to harden and preserve it... The other bones and stone tools had undoubtedly been planted in the pit and had been treated to match the dark brown color of the skull. *The lower jaw was that of a juvenile female orangutan. The place where the jaw would articulate with the skull had been broken off to hide the fact that it did not fit the skull.* The teeth of the mandible [lower jaw] were filed down to match the teeth of the upper jaw, and the canine tooth had been filed down to make it look heavily worn... The amazing thing about the Piltdown hoax is that at least twelve different people have been accused of perpetrating the fraud... what has been called *the most successful scientific hoax of all time*.[79] (emphasis added)

In **1915**, Sir Arthur Keith, Conservator of the Royal Medical College in England and President of the Royal Anthropological Institute of Great Britain and Ireland in the early 1900s, wrote the most definitive human evolution text of that era, *The Antiquity of Man*.[80] This 500+ page book prominently displayed a gold embossed skull of the Piltdown Man.

Over 100 pages of Arthur Keith's *The Antiquity of Man* book[81] is devoted to Piltdown Man, which was revealed as a fraud just two years before Keith died in 1955.[82] Keith placed so much trust in Piltdown Man as a "proof of evolution" that he called it: "one of the most remarkable discoveries of the twentieth century."[83] Boy was he wrong! But it was too late. He had convinced his readers that human evolution had scientific backing, when it never did.

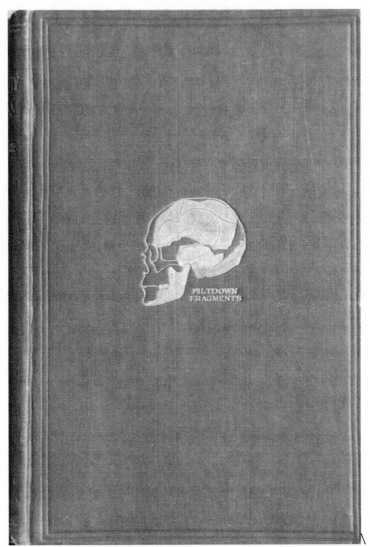

Figure 28. Sir Arthur Keith's Leading Human Evolution Book of the Early 1900s with Piltdown Man on the Cover[84] (Volume I, *The Antiquity of Man* by Sir Arthur Keith. Philadelphia: J.B. Lippincott Company, 1925. Second Edition, Sixth Impression. Illustrated).

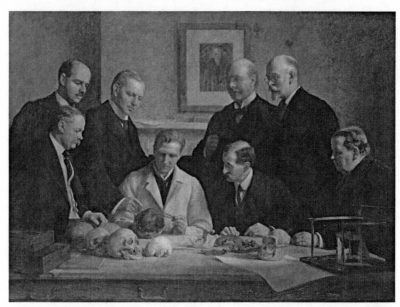

Figure 29. Group Portrait of the Piltdown Skull Examination. Back row (from left): F.O. Barlow, G. Elliot Smith, Charles Dawson, Arthur Smith Woodward. Front row: A.S. Underwood, Arthur Keith, W. P. Pycraft, and Ray Lankester. Painting by John Cooke, 1915. (Credit: Wikipedia).

For over forty years, Piltdown models were displayed around the world as proof of human evolution, and more than five hundred articles and memoirs are said to have been written about Piltdown man.[85] How did this fraud continue for so long before being exposed? Harvard paleontologist (and evolutionist) Stephen Gould suggests wishful thinking and cultural bias on the part of evolutionists was one reason.[86]

Figures 30–32 below show Piltdown's prominent place in leading human evolution "trees."

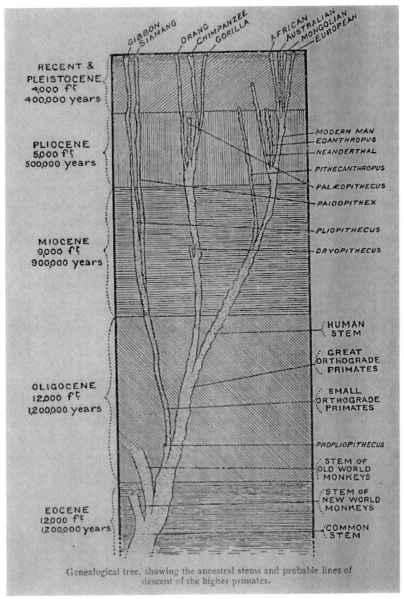

Figure 30. 1915 Human Evolution Ideas (Credit: Sir Arthur Keith, *The Antiquity of Man*. London: Williams & Norgate, 1915).

486   *SCIENCE*   [Vol. LXV, No. 1690

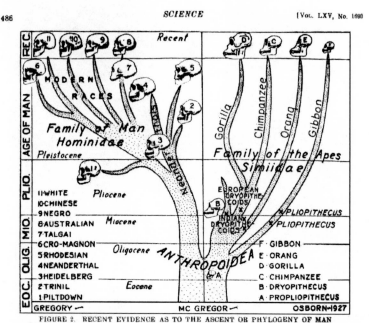

FIGURE 2. RECENT EVIDENCE AS TO THE ASCENT OR PHYLOGENY OF MAN

(LEFT) FAMILY OF MAN, *Hominidæ*, DIVIDING INTO THE NEANDERTHALOID (RIGHT) AND MODERN RACIAL (LEFT) STOCKS. PRESENT GEOLOGIC LOCATION OF THE PILTDOWN, HEIDELBERG, TRINIL, NEANDERTHAL AND RHODESIAN FOSSIL RACES (LEFT). (RIGHT) FAMILY OF THE APES, *Simiidæ*, INCLUDING THE PLIOCENE AND MIOCENE DRYOPITHECOIDS NEAREST THE ANCESTRAL STOCK OF THE *Anthropoidea*; ALSO THE LINES LEADING TO THE GORILLA, ORANG, CHIMPANZEE AND GIBBON. *Anthropoidea*—THE COMMON OLIGOCENE ANCESTORS OF THE *Hominidæ* (LEFT) AND OF THE *Simiidæ* (RIGHT).

Figure 31. 1927 Evolutionary Tree Showing Fraudulent Piltdown Man.[87]
Note Piltdown featured in the middle-left.

Figure 31 demonstrates Piltdown Man's prominent place in the supposed progression of human evolution. Piltdown models were displayed around the world as proof of human evolution for over 40 years, and illustrations including Piltdown Man in the chain of human evolution were used for decades in school textbooks.

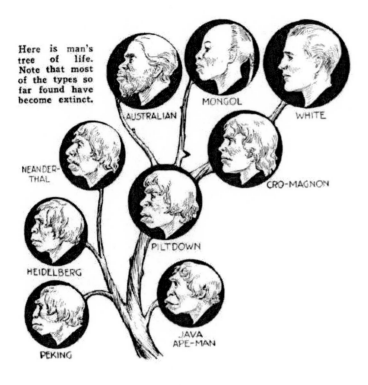

Here is man's tree of life. Note that most of the types so far found have become extinct.

Figure 32. 1931 Evolutionary Tree Showing Fraudulent Piltdown Man.[88] Piltdown Man, although completely faked, became standardized evidence for evolution.

NEBRASKA MAN

From **1917** to **1928**, yet another icon came on the scene as "certain proof" of human evolution. Geologist Harold Cook found a *single molar tooth* in Nebraska which later was named *Hesperopithecus haroldcooki*, or "Nebraska Man."

Figure 33. Nebraska Man (Credit: Wikipedia)

In **1922**, the head of the American Museum of Natural History (Henry Fairfield Osborn) proclaimed that the single molar found by Harold J. Cook in 1917 belonged to the first *pithecanthropoid* (ape-man) of the Americas, hence the name "western ape." The globally-distributed *Illustrated London News* broadcast British evolutionist Grafton Elliot Smith's receiving knighthood for his efforts in publicizing "Nebraska Man." This imaginative "reconstruction" of the tooth's owner is a club-carrying ape-man walking upright. It shows primitive tools, possibly domesticated animals, and a brutish bride gathering roots. An artist derived all this from a single tooth! In July 1925, the Nebraska Man tooth was used to prove man evolved from ape-like creatures in the Scopes "Monkey Trial" held in Dayton, Tennessee.

This all changed when excavations continued in 1927–1928 at the same place the tooth was found. These excavations revealed that the tooth belonged neither to man nor ape, but to a wild pig![89] Then, in 1972, living herds of this same pig were discovered in Paraguay, South America.[90] According to the late renowned creation scientist Duane T. Gish, "this is a case in which a scientist made a man out of a pig, and then the pig made a monkey out of the scientist!"[91]

SCOPES TRIAL

Next, the Scopes Trial of **1925** (Tennessee v. John Scopes) tested the state of Tennessee Butler Act, which prohibited the teaching of "any theory that denies the story of the Divine Creation of man as taught in the Bible, and to teach instead that man has descended from a lower order of animals." In other words, the Tennessee Butler Act made it illegal to teach human evolution in public school.

The Scopes Trial was one of the most famous trials of the 20th century, and public high school students still study it today—or at least watch the counterfactual black and white movie version titled *Inherit the Wind*. The famous criminal lawyer Clarence Darrow, known for believing that God was not knowable, represented John Scopes, a substitute high school teacher who was brought to trial for teaching evolution against State law. Three-time Democratic Presidential candidate and Christian William Jennings Bryan led the prosecution. The movie portrays him as a raving mad lunatic, but in real life he was calm, reasonable, and winsome. Scopes was found guilty under the Butler Act and was fined $100.

We bring up the Scopes Trial for three reasons. First, the case shows the growing tension in the creation-evolution debate and the extent to which each viewpoint was taught in school about one hundred years ago. Second, both Nebraska Man (a pig's tooth) and Piltdown Man (a complete forgery) were used as evidence to prove evolution at the Scopes Trial.[92] Third, legal battles regarding these issues resonate to this day. For example, removing the Ten Commandments, crosses, and nativity scenes from public spaces makes big news.

Progressing through the early- to mid-1900s, students continued to learn Biblically-based creation, even in public schools.[93] While some might find this difficult to believe because evolution theory is taught so widely in today's public schools, browsing public school textbooks from this earlier era easily confirms this fact. For example, in 1941 John Cretzinger investigated evolution teaching in fifty-four biological textbooks published between 1800 and 1933. He wrote, "The theory of Evolution was finally formulated by Charles Darwin in 1858, but it was destined to have little acceptance in secondary school books until after 1900 when the convincing evidence of Wallace and Haeckel made that theory acceptable as on the secondary science level."[94] Evolutionary theory was still only minimally represented in textbooks about one hundred years ago, with only token representations in junior high and high school texts.

In the 1950s, G.D. Skoog wrote, "... there was a continued increase in the emphasis on evolution in the textbooks from 1900 to 1950. This trend was reversed in the 1950s when the concept was deemphasized slightly."[95] A recent analysis of high school biology textbooks shows that emphasis on the topic of evolution decreased just before the 1925 Scopes Trial. The relative priority of evolution teaching retuned to pre-Scopes levels by 1935 and did not decrease significantly in the decades that followed.[96]

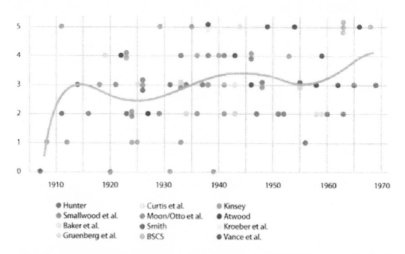

Figure 34. Relative Priority of the Topic of Evolution in Biology Textbooks 1907–1969.[97] The data in this image included 82 American high school biology textbooks published between 1907 and 1969. Each textbook was assigned a rating between 0 and 5 based on a qualitative assessment of the presentation of the topic of evolution.

Figure 34 shows a clear decline in the priority of the topic of evolution in the years ahead of Scopes trial in 1925, restoration of the topic to earlier levels by 1935, a secondary decline from about 1945 to 1955 and then a rise into the 1960s. A different analysis of ninety-three biology textbooks done by G.D. Skoog revealed:

> Analysis of the 93 biology textbooks revealed that prior to 1960, evolution was treated in a cursory and generally noncontroversial manner. However, there was a continued increase in the emphasis on evolution in the textbooks from 1900 to 1950. This trend was reversed in the 1950s when the concept was deemphasized slightly. In the 1960s the activities and influence of the Biological Sciences Curriculum Study (BSCS) resulted in several

textbooks that gave unprecedented emphasis to evolution. Accordingly, 51% of the total words written on the topics concerned with the study of evolution in the 83 textbooks published between 1900-1968 appeared in 17 textbooks published in the 1960s.[98]

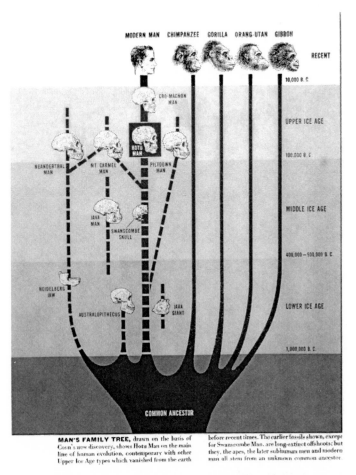

MAN'S FAMILY TREE, drawn on the basis of Coon's new discovery, shows Hotu Man on the main line of human evolution, contemporary with other Upper Ice Age types which vanished from the earth before recent times. The earlier fossils shown, except for Swanscombe Man, are long-extinct offshoots; but they, the apes, the later subhuman men and modern man all stem from an unknown common ancestor.

Figure 35. 1951 Life Magazine Evolutionary Tree[99] (still showing Java Man and Piltdown Man). This drawing shows a typical idea of human evolution in the 1950s, published in the well-known *Life Magazine* in 1951.

In **1959** a new fossil find filled a much-needed gap, since by then Nebraska and Piltdown frauds left nothing but a gaping hole that countless fossils should have filled if human evolution really happened. Enter *Zinjanthropus boisei*. National Geographic featured "Zinj," for short, as "Nutcracker Man" and framed it as "our real ancestor." Today, "virtually no evolutionist believes anymore that Zinj was our ancestor, but the images remain deep in millions of subconscious minds, reinforced by successive waves of other, often similarly temporary, "ape ancestor" images."[100] What happened? Further investigation revealed they were just extinct apes. Scientists have renamed them *Paranthropus*, and decided that they evolved alongside humans, not as our ancestors.

Next, in 1960 anthropologists uncovered remains from various locations at Olduvai Gorge in northern Tanzania and cobbled them together to make *Homo habilis*. *Homo habilis*, discussed in detail below, clearly does not fit in the line-up of human ancestry.

The concept of evolution became easier to believe in **1965** when Time-Life Books published the infamous "March of Progress" illustration in *Early Man*.[101] This book included a foldout section (shown in Figure 36) that displayed the sequence of figures drawn by Rudolph Zallinger.

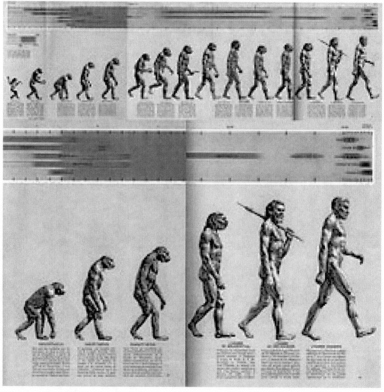

Figure 36. Zallinger's *March of Progress* (1965) (Credit: Wikipedia)

The year **1974** welcomed the famous "Lucy," a fossil form that bears the name *Australopithecus afarensis*. Lucy is arguably the most famous human evolution icon ever displayed in public school textbooks. Pictures and dioramas of Lucy inhabit countless museums and thousands of articles and dissertations. Lucy will be extensively discussed in the next section, where we expose details showing that it was merely an extinct ape.

A. afarensis   A. africanus   A. robustus   A. boisei   H. habilis   H. erectus   H. sapiens (archaic)   H. sapiens (Neandertal)   H. sapiens (modern)

Figure 37. *National Geographic* "March of Progress" (1985). Moving into the 1980s, this image provides an example of the current thinking about human evolution. (Credit: *National Geographic magazine*, 1985)

While Figure 37 was designed to show the alleged progression of "the evolution of running," it demonstrates the amazing imagination that artists have when taking scant fossil evidence and making them look increasingly human by lining them up side-by-side and altering their anatomies to fit the story. One such artist admitted: "I wanted to get a human soul into this ape-like face to indicate something about where she was headed."[102] Medical doctor Matthew Thomas wrote, "If today's police detectives obtained and interpreted evidence following these same principles/guidelines there would be chaos... yet we're supposed to accept this in science—paleontology—a field that seems to produce such abundant returns from such few fragments of fact![103]

EVOLUTION TEACHING IN TODAY'S PUBLIC SCHOOLS

Fast forward to today, where human evolution along with evolution theory in general, are taught as fact in public schools. The map shown in Figure 38 shows the creation-evolution teaching by state and school type (private/public). This map reveals that only two states (Louisiana and Tennessee) allow the Biblical view of Creation to be widely taught in public schools. The state of Texas has several charter schools that allow Creation curricula and nine states have private schools that accept tax-funded vouchers or scholarships that provide creation-based curricula.

Figure 38. Creation-Evolution Teaching by State.[104]

Amazingly, all states taught creation just one hundred years ago. It is even more shocking when considering the fact that about 70% of Americans profess Christianity,[105] and 46% of Americans believe that God created humans miraculously less than ten thousand years ago (see below). Wow—why do 96% of the states (48 of 50) teach "evolution as fact" in public schools, while 70% of America is "Christian" and 46% believe that God miraculously created humans rather recently? We offer some spiritual answers to this complex question below. Fortunately, the vast majority of homeschoolers in the U.S. use creation-based curriculum, and most private Christian schools use creation-based curricula that treats Genesis historically.

HISTORY TOUR WRAP-UP

While going through the 150-year "tour" through man's ideas of human origins, did you notice that the story changes substantially every few decades? Neanderthals were used to prove the "pre-human" myth from **1829 until the 1950s,** when they were shown to be human in almost every practical sense: burying their dead, making instruments, practicing burial rituals, using advanced tools, and even being buried alongside humans.[106]

Java Man fooled the world from **1891 to 1939**. Nebraska Man (a pig's tooth) filled the gap from **1917 until 1927**. Piltdown Man (a fraud) reigned from **1912 until 1953**. It seems like when one icon deceives a generation, a new one is introduced to save the day, and carry the evolutionary ideas for another generation.

Biblical creation, however, fits both "reality" and the fossil record much better. In reality, apes reproduce after their own kind and humans reproduce after theirs. And in the fossil record we see apes (including some extinct apes) and humans in a variety of shapes, and sizes. Why would you want to put your faith and understanding of our origins in a "science" that clearly changes its mind every twenty years? The Biblical position has fit the facts since the beginning and has never changed.

WHY IS EVOLUTION TAUGHT IN PUBLIC SCHOOLS?

The following chart shows the percentage of all Americans who hold the creationist view that God created humans in their present form within the last ten thousand years. This has been the predominant view since the question has been tracked by the Gallup Poll for the past thirty years. About one-third of Americans believe that humans evolved, but with God's guidance; 15% say humans evolved, but that God had no part in the process.

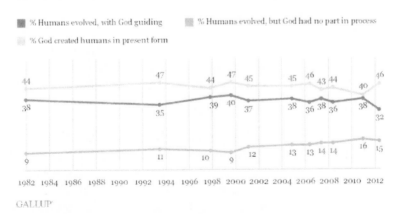

Figure 39. American Positions on Human Origins: 1982 to 2012[107]

If the majority position on human origins is "Divine Creation less than 10,000 years ago," then why do the majority of secular institutions today teach that humans evolved, a position which is held by only 15% of the U.S. population? Some might answer this question by stating, "Separation of church and state" or by offering some plausible political explanations. Some might say that "apathetic Christians" are to blame. While these and other explanations might seem to fit, we offer a different reason—a *spiritual* reason.

The Bible gives the purpose behind a strong delusion that will arise. "They perish because they refused to love the truth and so be saved. For this reason, God sends them a powerful delusion so that they will believe the lie and so that all will be condemned who have not believed the truth but have delighted in wickedness" (2 Thessalonians 2:10–12). Could the "powerful delusion" be evolution and "the lie" be that God does not exist? This delusion is sent to those who chose not to believe the Gospel of Christ. It takes more faith to believe in evolution than Creation, and Romans 1 makes it clear that everyone will be held accountable to the Creator because of His Creation:

> The wrath of God is being revealed from heaven against all the godlessness and wickedness of people, who suppress the truth by their wickedness, since what may be known about God is plain to them, because God has made it plain to them. For since the creation of the world God's invisible qualities—his eternal power and divine nature—have been clearly seen, being understood from what has been made, so that people are without excuse. (Romans 1:18–20)

This passage continues to explain that God allows those who disregard and reject Him to earn their own condemnation. This process is happening in America at a record pace, showing that some prophesies in Scripture are increasingly relevant today:

> Above all, you must understand that in the last days scoffers will come, scoffing and following their own evil desires. They will say, "Where is this 'coming' he promised? Ever since our ancestors died, everything goes on as it has since the beginning of creation." But they deliberately forget that long ago by God's word the heavens came into being and the earth was formed out of water and by water. By these waters also the world of that time was deluged and destroyed. (2 Peter 3:3–6)

Peter warns that these scoffers would "willfully forget" about the Biblical creation account and the catastrophe of the Flood. Just how does someone "willfully forget" about something? By *teaching the opposite*— which is exactly what's happening in today's public schools. Those who deny Creation replace the worldwide Flood with a local flood coupled with long-age geology, just as Peter foretold. Indeed, the mission that was started by Charles Lyell in 1830 to "free science from Moses"[108] (meaning the Genesis creation and Flood accounts) has made incredible progress.

Each person has a choice. We can accept and believe the truth of Jesus Christ as presented in the Scriptures: that His death paid our sin debt and that His resurrection paved the way to everlasting life for believers. "This is love for God: to obey His commands" (1 John 5:3). He commands everyone everywhere to turn from their sins and go to Him. Conversely, to know the truth and not obey it earns the wrath of God: "The wrath of God is being revealed from heaven against all the godlessness and wickedness of men who suppress the truth by their wickedness" (Romans 1:18). Frankly, there is no more dangerous condition for man than to know the truth and refuse to obey it. To do so is to harden the heart and make God's condemnation sure.

"Professing themselves to be wise they became fools."
— Romans 1:22

"Why is not every geological formation and every stratum full of such intermediate links? Geology assuredly does not reveal any such finely graduated organic chain; and this is the most obvious and serious objection which can be urged against the theory."
— Charles Darwin, Origin of Species, 1872.

"There is a popular image of human evolution that you'll find all over the place ... On the left of the picture there's an ape ... On the right, a man ... Between the two is a succession of figures that become ever more like humans ... Our progress from ape to human looks so smooth, so tidy. It's such a beguiling image that even the experts are loath to let it go. But it is an illusion."
— Bernard Wood, Professor of Human Origins, George Washington University[109]

"Fossil evidence of human evolutionary history is fragmentary and open to various interpretations. Fossil evidence of chimpanzee evolution is absent altogether."
– Henry Gee, Senior Editor of Nature Magazine[110]

"Modern apes, for instance, seem to have sprung out of nowhere. They have no yesterday, no fossil record. And the true origin of modern humans—of upright, naked, tool-making, big-brained beings - is, if we are to be honest with ourselves, an equally mysterious matter."
– Dr. Lyall Watson, Anthropologist[111]

CHAPTER 3

Typical Ape-to-Human Progression in Public School Textbooks

Daniel A. Biddle, Ph.D. & David Bisbee

The four "stages" of human evolution typically presented in Sixth Grade Social Studies (World History) classes looks like this:

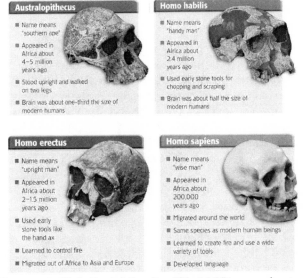

Figure 40. Holt Social Studies World History: Ancient Civilizations[112](Credit: Holt Social Studies *World History: Ancient Civilizations*, Holt, pages 24-35, 2006)

Next we'll review each of these "ape-to-human" icons one at a time, starting first with "Lucy."

AUSTRALOPITHECUS AFARENSIS ("LUCY")

In 1974, Donald Johanson discovered a fossil in Ethiopia, Africa that he declared was the "missing link" between man and ape. The fossil was nicknamed "Lucy" and was given the scientific name *Australopithecus afarensis*. Australopithecus simply means "southern ape." Southern ape is a very appropriate name because, as you'll learn below, Lucy was just that— an ape!

Although public school textbooks often state that Lucy was our ancestor and feature human-like drawings of her, the fossil evidence tells quite a different story. Now, after forty years of research on Lucy and other Australopithecine fossils, here is what scientists have found (Note: because more Australopithecine fossils have been found since Lucy, some of the references below refer to *Australopithecines* in general):

- **Entire Skeleton:** Even though many of the first reports that came out after Lucy was discovered stated that Lucy's skeleton was "40% complete,"[113] Lucy's discoverer clarified this in a book published twenty-two years after[114] Lucy was found stating: "Lucy's skeleton consists of some 47 out of 207 bones, including parts of upper and lower limbs, the backbone, ribs and the pelvis. With the exception of the mandible [lower jaw] the skull is represented only by five vault fragments, and most of the hand and foot bones are missing." This computes to actually **22.8%** of the complete skeleton (47 ÷ 206), not "about 40%." Generations of artists have drawn Lucy with human feet even though the fossil lacked both hand and foot bones. Frustratingly for those who care about truth, these illustrations continue to ignore subsequent finds, revealing that *Australopithecines* had curved ape fingers and grasping ape feet.
- **Skull:** Even though only a few fragments of Lucy's skull were found, they revealed that her skull was about the same size as a chimpanze. As Donald Johanson himself said, "Her skull was almost entirely missing. So knowing the exact size of Lucy's brain was the crucial bit of missing evidence. But from the few skull fragments we had, it looked surprisingly small."[115] Later estimates reveal that Lucy's brain was just one third the size of a human brain, which makes Lucy's brain the same size as the average chimpanze brain.[116] Sir Solly Zuckerman, chief scientific advisor to the British government,

said that the *"Australopithecine* skull is in fact so overwhelmingly ape-like, as opposed to human that the contrary position could be equated to an assertion that black is white."[117]

- **Height:** Lucy was about 3.5 feet tall (and most other *Australopithecine* fossils found since are similar in height).

- **Walking Upright:** Even evolutionists strongly disagree over whether or not Lucy walked upright like humans.[118] Lucy's hip was found broken and was reconstructed, so it's difficult to tell how she (and other *Australopithecines*) moved. Her bones seemed to show that she was a "real swinger... based on anatomical data, *Australopithecines* must have been arboreal [tree-dwelling]...Lucy's pelvis shows a flare that is better suited for climbing than for walking."[119] Most likely, Australopithecine apes could walk in their own unique way— unlike chimps or humans.

- **Fingers and Limbs:** Other examples of *Australopithecine* apes had curved fingers and ape-like limb proportions that point toward her kind as living in trees, so we can assume the same was true of Lucy.[120]

- **Locking Wrists:** Lucy had locking wrists like quadruped apes, not like humans.[121] This was even reported in the *San Diego Union Tribune*: "A chance discovery made by looking at a cast of the bones of 'Lucy,' the most famous fossil of *Australopithecus afarensis*, shows her wrist was stiff, like a chimpanzee's, Brian Richmond and David Strait of George Washington University in Washington, D.C., reported. This suggests that her ancestors walked on their knuckles."[122] Another study revealed: "Measurements of the shape of wristbones (distal radius) showed that Lucy's type were knuckle walkers, similar to gorillas."[123]

- **Teeth:** The wear on Lucy's teeth indicate she ate tree fruit.[124] Penn State University professor of anthropology and biology Alan Walker has studied paleontological fossils to learn how to reconstruct their ancient diets. In speaking of Alan Walker's material, Johanson noted: "Dr. Alan Walker of Johns Hopkins has recently concluded that the polishing effect he finds on the teeth of robust [thick-boned] *Australopithecines* and modern chimpanzees indicates that *Australopithecines*, like chimps, were fruit eaters.... If they were primarily fruit eaters, as Walker's examination of their teeth suggests they were, then our picture of them, and of the evolutionary path they took, is wrong."[125]

- **Ribs:** Lucy's rib cage is not shaped like a human's, but was cone shaped like an ape's.[126] Peter Schmid, a paleontologist at the Anthropological Institute in Zurich, Switzerland, studied a replica of Lucy and noted: "When I started to put the skeleton together, I expected it to look human. Everyone had talked about Lucy being very modern. Very human. So I was surprised by what I saw. I noticed that the ribs were more round in cross section. More like what you see in apes. Human ribs are flatter in cross section. But the shape of the ribcage itself was the biggest surprise of all. The human ribcage is barrel shaped. And I just couldn't get Lucy's ribs to fit this kind of shape. But I could get them to make a conical-shaped ribcage, like what you see in apes."[127]

- **Ears:** Earlier in this book we learned that an animal's semicircular canals help reveal its identity. After extensive research, it has been concluded that the semicircular canals of *Australopithecines* resemble an ape's, not a human's or a transitional creature's.[128]

- **Gender:** A great deal of debate has emerged even over Lucy's gender, with some scientists arguing that the evidence shows she was actually a male! Articles with catchy titles have emerged such as "Lucy or Lucifer? [129] and more recently, "Lucy or Brucey?"[130]

- **Toes:** The toe bones of *Australopithecines* were long and curved, even by ape standards.[131] Their fossils thus give no evidence that they walked like humans. Instead they show strong evidence that they did not. It is because of these recent findings that leading experts in *Australopithecine* fossils conclude that Lucy and other *Australopithecines* are *extinct ape-like creatures*:

  - Dr. Charles Oxnard (professor of anatomy) wrote, "The *Australopithecines* known over the last several decades ... are now irrevocably removed from a place in the evolution of human bipedalism...All this should make us wonder about the usual presentation of human evolution in introductory textbooks."[132]

  - Dr. Solly Zuckerman heads the Department of Anatomy of the University of Birmingham in England and is a scientific adviser to the highest level of the British government. He studied Australopithecus fossils for 15 years with a team of scientists and concluded, "They are just apes."[133]

  - Dr. Wray Herbert admits that his fellow paleoanthropologists "compare the pygmy chimpanzee to 'Lucy,' one of the oldest hominid fossils known, and finds the similarities striking. They are almost identical in body size, in stature and in brain size."[134]

  - Dr. Albert W. Mehlert said, "the evidence... makes it overwhelmingly likely that Lucy was no more than a variety of pygmy chimpanzee, and walked the same way (awkwardly upright on occasions, but mostly quadrupedal). The 'evidence' for the alleged transformation from ape to man is extremely unconvincing."[135]

  - Marvin Lubenow, Creation researcher and author of the book *Bones of Contention,* wrote, "There are no fossils of Australopithecus or of any other primate stock in the proper time period to serve as evolutionary ancestors to humans. *As far as we can tell from the fossil record, when humans first appear in the fossil record they are already human*[136] (emphasis added).

- Drs. DeWitt Steele and Gregory Parker concluded: "Australopithecus can probably be dismissed as a type of extinct chimpanzee."[137]

In reality, these ape-like creatures' remains occur in small-scale deposits that rest on top of broadly extending flood deposits. They were probably fossilized after Noah's Flood, during the Ice Age, when tremendous rains buried Ice Age creatures.[138] Donald Johanson, the discoverer of Lucy, admits: "The rapid burial of bones at Hadar, particularly those of the 'First Family,' are related to a geological catastrophe suggesting, perhaps, a flash flood. Bones are fragmented and scattered because individuals fell into a river, or were washed into a river, rapidly transported, broken up, and scattered. These are all products of a depositional process."[139]

Despite these recent findings, Lucy continues to be displayed more human-like than her fossils would justify. Some examples of these exaggarations at public museums and in textbooks are below. First, let's look at what they actually found:

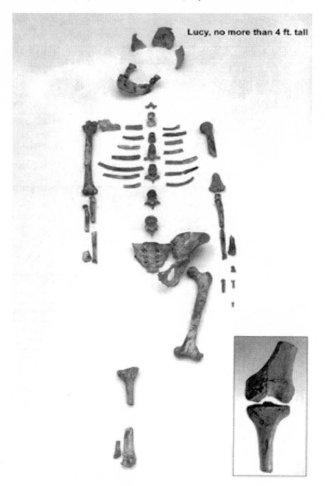

Figure 41. Actual Lucy Fossil
(Credit: Answers in Genesis Presentation Library)

Before viewing some renditions that superimpose human characteristics on Lucy, let's start with what she probably looked like.

Figure 42. What Lucy Most Likely Looked Like
(Credit: Answers in Genesis Presentation Library)

Next, let's look at how Lucy is represented at public exhibits, such as those found at the St. Louis Zoo and Denver Museum of Nature and Science.

Figure 43. Lucy at Public Exhibits (Zoos and Museums). Lucy at the St. Louis Zoo (left) (Credit: Answers in Genesis) and at the Denver Museum of Nature and Science. (Credit: Brian Thomas)

Most Lucy reproductions show her with white sclera (eyeballs), even though 100% of all apes alive today have dark eyes. Do you think this was done to make her look more human-like?

Figure 44. Lucy with White Sclera (Eyeballs). Like similar Hollywood characters, this imaginative version of Lucy presents it with human eyes, though eyes don't fossilize. (Credit: Wikipedia)

Now let's view how Lucy is typically represented in public school textbooks:

Figure 45. Lucy in Public School Textbooks [Credit: Australopithecus afarensis (*History Alive! The Ancient World* (Palo Alto, CA: Teachers Curriculum Institute, 2004)].

Next, let's take a look at where 100% of the Australopithecus fossils have been found (see circles in Figure 46).

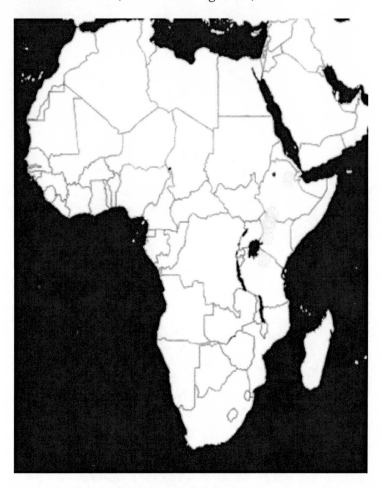

Figure 46. Map Showing where Australopithecus Fossils have been Found [140] (Credit: *www.fossilworks.com*)

Here is one interesting fact that you won't likely learn about from school textbooks: At the specific site where Lucy was found, eighty-seven other animal types were discovered. This was a wide collection that included just about every animal you would expect to see residing with ape-like creatures, including elephants, rhinoceros, hippopotamus, antelope, and numerous other African-native animals. In fact, this specific area (the Hadar Valley formation) has yielded nearly six thousand specimens representing as many as four thousand different animals.[141] It certainly makes sense that apes in Lucy's day were living with similar creatures in a similar habitat as ape-like creatures today!

Now if Lucy's fossil looks like an ape, if she lived with other apes, if she lived in an environment like apes today, and if she lived with eighty-seven other animal types that live around apes, what do you think she was?

HOW MANY AUSTRALOPITHECUS AFARENSIS FOSSILS HAVE BEEN FOUND?

An online research tool known as the Global Biodiversity Information Facility (GBIF) tabulates various fossil specimens found around the world. This free tool provides a single point of access to more than five hundred million records, shared freely by hundreds of institutions worldwide, making it the biggest biodiversity database on the internet, with information regarding more than 1.5 million species.[142]

Using GBIF to research *Australopithecus afarensis* fossils reveals a total of forty-seven "occurrences" (individual findings or dig sites where multiple specimens have been found). Browsing through these "occurrences" reveals just how limited the findings are for this species. The biggest occurrence is called the "First Family" where 260 bones and bone pieces were found representing between thirteen to seventeen creatures.[143] The vast majority of the bones were found within the top few feet of the surface, indicating they likely died at the same time.[144]

Some recent estimates place the total count of *Australopithecus afarensis* fossils at only 362 fragments,[145] which likely represents only a few dozen individual creatures. With only 260 of these fragments coming from one "family," and another forty-seven from Lucy, one wonders where all the leftovers are from supposedly millions of years of Lucy populations. If human evolution really occurred as the textbooks state, wouldn't we expect to find, as Charles Darwin stated, "innumerable transitional forms"[146] and "every geological formation full of intermediate links"? Clearly, as Darwin himself admitted, "Geology does not reveal any such finely-graduated chain; and this is the most obvious and serious objection against the theory."[147]

Even though earth layers have revealed precious few Australopithecine fossils, they reveal all we need to know: Lucy was an ape. Even the most recent Human Family Tree from the Smithsonian Institute shows that *Australopithecines* are not even on the same "branch" of the tree that includes *Homo habilis* and *Homo erectus*!

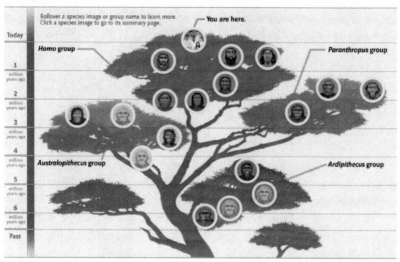

Figure 47. Smithsonian Institute Human Family Tree[148]

HOMO HABILIS

Homo habilis or "handy man" is often shown in public school textbooks as a "transitional" form between apes and humans. Textbooks state that this evolutionary ancestor supposedly lived around 1.4–2.4 million years ago[149] and was one of the "stepping stones" in the line of human evolution. In reality, *Homo habilis* is not just one fossil, but rather a *very* small collection of fossils that have been the center of intense controversy and confusion for decades. With so many now agreeing that Lucy was almost all ape, and with so many agreeing that the other species in genus Homo, including Neanderthals, are modern man, evolutionists are desperate for a genuine link between apes and man. Frankly, if *Homo habilis* fails to connect apes to humans, then human evolution fails with it.

The name *Homo habilis* was officially given to a set of fossils that were discovered by a team led by scientists Louis and Mary Leakey between 1960 and 1963 at Olduvai Gorge in Tanzania. In 1964, this team announced *Homo habilis* as a "new human ancestor." The original fossils were said to be 1.8 million years old and consisted of scattered skull parts, hand bones, and foot bones from four young specimens. According to

Louis Leakey, the foot bones showed signs that *Homo habilis* may have been able to walk upright on two feet, and the hand bones indicated they were skillful with their hands. However, since these bones were not found next to the skull fragments, there was no way to be sure that they belonged to the same creature. Some evolutionary scientists even believe that the *Homo habilis* fossils were just a mixture of *Australopithecine* (ape) and *Homo erectus* (human) fossils—not a new species at all and certainly not a missing link.[150]

The Leakeys also found some primitive stone tools at the site. Originally scientists claimed these tools belonged to another supposed missing link known as *Zinjanthropus,* which turned out to be just an ape. Louis Leakey claimed the tools were used by their newfound individuals. This was the reason for naming these fossils "*Homo habilis*" or handy man. But because we weren't there to observe these creatures we don't know if the creatures used the tools, or if the tools were used on them!

In 1986, Tim White and Don Johanson discovered a partial adult skeleton. Since the fossil was discovered in Olduvai Gorge, it was designated "Olduvai Hominid 62" and was dated (by evolutionists) at 1.8 million years old. Because the skull and teeth were similar to the original *Homo habilis* fossils found in 1964, the new fossil was said to belong to the same species. This presented three big problems for evolutionists:

1. The body of Olduvai Hominid 62 was rather ape-like and even smaller than the famous Australopithecine fossil known as Lucy.[151] Since Lucy was about 3.5 feet tall, and *Homo erectus* individuals grew to be about six feet, Olduvai Hominid 62 should have been somewhere in between them if it truly links the two.
2. Since the body of Olduvai Hominid 62 was ape-like, it seemed to support the belief that the original *Homo habilis* fossils found in 1964 were actually a mixture of Australopithecine parts and human bones, most notably human hands and feet.
3. If the fossilized hand and feet bones found in 1964 were actually human, then the tools found at that site were probably used by people living there—not by ape-like people, or people-like apes, but the descendants of Adam and Eve.

Despite the bold statements made about *Homo habilis* in many school textbooks, paleoanthropologists are still trying to make sense out of this odd collection of fragments. Here is how evolutionist Richard Leakey described the problem: "Of the several dozen specimens that have been said at one time or another to belong to this species, at least half of them don't. But there is no consensus as to which 50% should be excluded. No one anthropologist's 50% is quite the same as another's."[152] The same could be said of every proposed missing link. For every evolutionist who asserts that a particular fossil belongs in human ancestry, another one counter-asserts that it evolved parallel to the unknown evolutionary ancestors of man. What a mess.

Some studies have revealed that the ears of the *Homo habilis* specimens studied prove they were just apes. Of course, these results don't cover the fossil bits attributed to this name that actually belong to another. Anatomy specialists Fred Spoor, Bernard Wood, and Frans Zonneveld compared the semicircular canals in the inner ear of humans and apes, including several Australopithecus and *Homo habilis* specimens. Because the semicircular canals are involved in maintaining balance, studying them can reveal whether an animal was inclined to walk upright or on all fours. Their study concluded: "Among the fossil hominids [apes or humans] the earliest species to demonstrate the modern human morphology is *Homo erectus*. In contrast, the semi-circular canal dimensions in crania from southern Africa attributed to *Australopithecus* and *Paranthropus* resemble those of the extant great apes."[153] The authors wrote that *Homo habilis* "relied less on bipedal behavior than the *Australopithecines*," meaning that the *Homo habilis* specimen was even more ape-like than the Australopithecus samples. They concluded that the *Homo habilis* specimen they studied "represents an unlikely intermediate between the morphologies seen in the *Australopithecines* and *Homo erectus*."[154] In other words—the *Homo habilis* is a "mixed bag" classification that includes some ape bones and some human bones. In summary, their study resulted in two very important findings:

1. These *Homo habilis* fossils do not actually belong to the "human" group, but rather to an ape category, and probably *Australopithecus*.
2. Both *Homo habilis* and *Australopithecus* walked stooped over like an ape and not upright like a man.

The claim that *Australopithecus and Homo habilis* walked upright was disproved by inner ear analyses carried out by Fred Spoor. He and his team compared the centers of balances in the inner ears, and showed that both moved in a similar way to apes of our own time.

Figure 48. Semicircular Canals[155]

So was *Homo habilis* really our ancestor? Even evolutionists disagree. Dr. Bernard Wood of George Washington University, an expert on evolutionary "trees," suggests that none of the *Homo habilis* fossils represent human ancestors. He wrote, "The diverse group of fossils from 1 million years or so ago, known as *Homo habilis*, may be more properly recognized as *Australopithecines*."[156]

In a more recent article titled, "Human evolution: Fifty years after *Homo habilis*," Dr. Wood summarizes more than one-half of a century of research on *Homo habilis* by concluding that:

> Although *Homo habilis* is generally larger than Australopithecus *africanus*, its teeth and jaws have the same proportions. What little evidence there is about its body shape, hands and feet suggest that *Homo habilis* would be a much better climber than undisputed human ancestors. So, if *Homo habilis* is added to Homo, the genus has an incoherent mishmash of features. Others disagree, but I think you have to cherry-pick the data to come to any other conclusion. My sense is that handy man should belong to its own genus, neither australopith nor human.[157]

Although evolutionists keep trying to convince themselves (and others) that humans evolved from ape-like creatures, interpretations of the fossil record have been filled with mistakes, fraud, and fantasy, with almost every major pronouncement denounced by another expert. Why don't textbooks tell these truths? Perhaps before even examining the evidence, textbook writers reject the truth that we were created by God on day six of creation week. Since the beginning, humans have always been humans and apes have always been apes. And since Adam and Eve sinned, humans have worked extra hard to ignore our Creator.

HOMO ERECTUS

Homo erectus means "erect or upright man." Typically, school textbooks claim that *Homo erectus* fossils fill the gap between *Australopithecines* (apes) and both Neanderthals and modern humans.

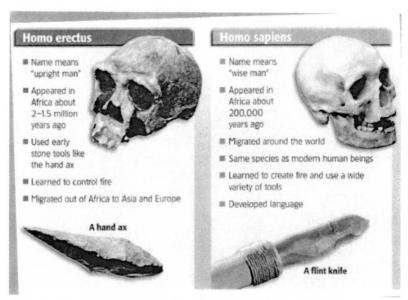

Figure 49. Homo erectus in School Textbooks [Credit: Holt Social Studies *World History: Ancient Civilizations* (Holt, 2006, pages 24-35)]

Here is an example of what a middle school textbook (Holt, 2006) teaches about *Homo erectus*:

- The name *Homo erectus* means "upright man."
- Scientists agree that *Homo erectus* was not fully human and was the evolutionary link between *Homo habilis* and *Homo sapiens.*
- *Homo erectus* first "appeared" in Africa 2 to 1.5 million years ago and migrated to Asia and Europe.
- *Homo erectus* used early stone tools and learned to control fire.

Although school textbooks (like the one shown above) often teach that we evolved from primitive, sub-human ancestors known as *Homo erectus,*[158] the growing creationist (and evolutionist) view is that *Homo erectus* and all *Homo sapien* forms should be considered not as separate species but as a single human species that represent a wide range of diversity. In the Biblical Creation view, there was no evolution from apes, nor was there any "ascent" from an inferior human type to a more advanced kind.[159] A total of about 280 *Homo erectus* fossils have been found to date.[160] They include bones, bone fragments, and teeth.

Some evolutionists claim that the size of the skulls helps determine how far along a creature is in its journey towards becoming human. The skulls designated *Homo erectus* fall within the cranial capacity range of modern humans (700 cc to 2,100 cc).[161] Marvin Lubenow, an expert on human fossils, comments: "My own conclusion is that *Homo erectus* and Neanderthal are actually the same: *Homo erectus* is the lower end, with regard to size, of a continuum that includes *Homo erectus,* early *Homo sapiens,* [who looked just like people today] and Neanderthal. The range of cranial capacities for fossil humans is in line with the range of cranial capacities for modern humans."[162]

One study compared modern humans to *Homo erectus* fossils including Java Man, Peking Man, and East African Man. What they found was a big surprise to many evolutionists: A group of 202 modern day Australian aborigines share an astonishing 14 of the 17 *Homo erectus* traits.[163] The most recent evidence indicates that only a handful of features distinguish these two presumed species of man, and even these are doubtful.[164] Nobody should doubt the fully human status of Australian aborigines, so why doubt the fully human status of most fossils designated as *Homo erectus?*

Although *Homo erectus* is supposed to represent an evolutionary link between *Homo habilis* and *Homo sapiens,* its fossils occur throughout most layers thought to contain human evolution remains. The dates evolutionists assigned to these fossils show that *Homo erectus* lived during the **same time periods** as both *Homo habilis* (a category that includes a mixture of both Australopithecine and human fossils) and modern humans.[165] How could *Homo erectus* be an evolutionary link if they lived at the same time?

Table 1. Secular Homo erectus dates overlap with modern humans[166]

Homo erectus Fossil Name	Date Assigned by Evolutionists
Swartkrans SK-15, 18a and 18b	1.8 million years
KNM-WT15000 Kenya	1.6 million years
Kow Swamp Fossils	9,500 years
Cossack Skull	6,500 years
Mossgiel Individual	6,000 years

Many of the artifacts found with *Homo erectus* fossils show that they intelligently used tools, built shelters, controlled fire and even carved quartzite rocks into human figurines.[167] During the early 1800s, many Native Americans lived in a similar manner. In other words, although they were not as technologically advanced as some other cultures, they were fully human. The bones in the *Homo erectus* classification are really nothing more than an example of human variability. The next time you visit a public place, take a good look at the people around you (politely, of course). You'll see humans come in a wide variety of shapes and sizes.

HOMO SAPIENS

Homo sapiens means "wise man" in Latin and is the scientific name for mankind. The human genus "Homo" includes Neanderthals and *Homo sapiens sapiens* ("wise, wise man"). Some sources show Neanderthals (Homo *neanderthalensis*) as a subspecies of modern man by accepting the name *Homo sapiens neanderthalensis*.[168] While evolutionary thinkers search for tiny differences on which to base their pre-judgment of ape ancestry, creation-based thinking sees fossil and modern variations as expressions of wide genetic potential that God built into Adam.

School textbooks often place "stone-age men" and "cavemen"

into the *Homo sapiens* category. Even the term "cavemen" is somewhat misleading because it assumes that ape-like men had not yet evolved enough intelligence to construct homes. However, throughout history people have lived in caves wherever caves exist—even to very modern times. Sometimes their cave homes were permanent, sometimes they were temporary, and sometimes people simply found temporary shelter or buried their dead in caves.

The term caveman, however, typically refers to people who lived before or during the Ice Age. Five groups fit this definition: Neanderthals, early *Homo sapiens* (Cro-Magnon man), *Homo erectus*, Denisovans, and Homo *floresiensis*. The latter two groups were recently added.[169] Researchers discovered these remains in caves. Who were these people?

A Biblical view on cavemen is simple: they were people who lived soon after the Flood, and they found temporary shelter in the caves that formed in the rock layers laid down by the Flood. Perhaps some cave-dwellers represented those who first scattered around the world from the Tower of Babel dispersal that Genesis chapter 11 describes. They sought caves as temporary and sometimes permanent shelters, especially during the post-Flood Ice Age.

According to Scripture, humans have been bright, innovative, and capable from the very beginning. We have seen the science of archaeology confirm this, as even cave-living humans left behind well-crafted tools. According to Genesis chapter 4, fifth generation humans like Tubal-Cain worked with metals including copper and iron. People were gardening, farming, working with different types of metal and even building cities before the Flood. Cain was a tiller of the ground. (Gen. 4:2). Later in Cain's life he built a city. Cain's eighth generation Jubal "was the father of all those who play the harp and flute."

After the Flood, much of this technology and know-how was lost, especially after people scattered around the world from the Tower of Babel dispersion. Let's take a closer look at the two most common "cavemen" described in public school textbooks to see which expectation their remains most closely match: that of less-than-human evolutionary ancestors or fully human early wanderers.

NEANDERTHAL MAN[170]

Neanderthal man was named after the Neander Valley near Dusseldorf in West Germany where the first fossils were found in 1856.

It gained its name because of the frequent visits by hymn writer Joachem Neander + *tal*, or *thal* in Old German, meaning "valley." Just as "Thomas" is pronounced "Tomas," so we pronounce "Neanderthal" as "Neandertal." Confusingly, experts use either spelling. The story of how evolutionists have classified Neanderthal from true man to "missing link" and then to variant forms of modern humans is as interesting as the people themselves.

Originally, "when the first Neanderthal was discovered in 1856, even 'Darwin's bulldog,' Thomas Henry Huxley, recognized that it was fully human and not an evolutionary ancestor."[171] Nevertheless, evolutionary bias helped anatomist William King reinterpret the fossils, concluding they were a separate, primitive species of man called *Homo neanderthalensis*. This designation easily fit the assertion that modern humans evolved from Neanderthals. More and better evidence, including burial sites that held Neanderthals and modern men in the same tombs, forced some evolutionists to change its name in 1964.

Today, with over two hundred known specimens representing more than forty discovery sites in Europe, Asia, and Africa, "Neanderthal fossils are the most plentiful in the world [of paleoanthropology]."[172] In recent decades this mound of data has testified to the fact that, "while the Neanderthals may not have been as culturally sophisticated as the people who followed . . . the Neanderthal people were not primitive but the most highly specialized of all the humans of the past."[173] "Evolutionists now admit that the Neanderthals were 100% human; they are classified as *Homo sapiens neanderthalensis*, designating them as a [subspecies] variety of modern humans."[174] Their skeletons were a bit thicker in places than most modern humans. They were up to 30% larger in body mass and had more than 13% larger brain volumes.

However, "the strongest evidence that Neanderthals were fully human and of our species is that, at four sites [3 in Israel and 1 in Croatia], Neanderthals and modern humans were buried together," indicating that "they lived together, worked together, intermarried, and were accepted as members of the same family, clan, and community" since generational "reproduction is on the species level."[175] Neanderthal burials include jewelry and purses, showing they had nothing to do with any ape-kind. Strikingly, the Neanderthal burial practice of using caves as family burial grounds or tribal cemeteries exactly parallels that of the post-Babel patriarchs of Genesis, for example Abraham (Genesis 23:17–20), Isaac (Genesis 25:7–11), and Jacob (Genesis 49:29–32.)

The lifespan of the Neanderthal people also looks astonishingly similar to the lifespan of those living in the post-Flood generations including Peleg (Genesis. 11:12–17). Using recent dental studies and digitized x-rays, computer-generated projections of orthodontic patients have illustrated the continuing growth of their craniofacial bones. These show a Neanderthal-like profile of the skull as the patient advances into their 300th, 400th, and even 500th year of simulated life.[176] Career dentist Dr. Cuozzo analyzed teeth and jaw development in children. He wrote, "studies on aging reveal that the older we get, the more our faces begin to look like those of Neanderthal man. The most accurate assumption that can be made about these strange-looking skeletons that are not old enough to be fossilized is that they have been alive long enough for their bones to change into those shapes—they are skeletons of patriarchs who lived hundreds of years, but have only been dead for thousands of years, not millions!"[177]

Creation researchers have been saying for decades that Neanderthal man was wholly human, with no hint of a single evolutionary transitional feature. Neanderthal DNA sequences published in 2010 confirmed this, and showed that certain people groups today share bits of Neanderthal-specific DNA sequences.[178]

CRO-MAGNON MAN

Cro-Magnon Man is known as the "big hole man" in the French dialect local to the initial 1868 discovery site, a cave in the Dordogne area of Les Eyzies in southwest France. Once regarded as our most recent evolutionary ancestors on the "ape-to-man" illustrations, "evolutionists now admit that Cro-Magnons were modern humans. Cro-Magnons are classified as Homo *sapiens sapiens* [wise, wise man'], the same classification assigned humans today."[179] Creation writer Vance Ferrell echoed this consensus when he wrote, "the Cro-Magnons were normal people, not monkeys; and they provide no evidence of a transition from ape to man."[180] With interests ranging from stone tools, fishhooks, and spears to more sublime activities like astronomy, art, and the afterlife, "every kind of evidence that we have a right to expect from the fossil and archeological record indicates that the Cro-Magnon and Neanderthal peoples were humans in the same ways that we are human."[181]

Contrary to popular belief, most Cro-Magnon people used caves for rituals, not residences. In addition, authenticated etchings on the cave walls at Minetada, Spain in 1915, and La Marche, central France (1937), depict Cro-Magnon men with clipped and groomed beards while the women display dresses and elegant hair styles.[182] Advanced not only in manner but also in the way they looked: "the Cro-Magnons were truly human, possibly of a noble bearing. Some were over six feet tall, with a cranial volume somewhat larger (by 200cc–400cc) than that of man today."[183] Brain size should not be exclusively used to judge whether or not a given specimen was human or not, but it can, in combination with other skull features, add its testimony. In any case, just as with Neanderthal man, Cro Magnon men were wholly human. Why do illustrations of human evolution show them walking up behind modern men if they showed no real differences after all?

CHAPTER 4

What about the Different "Races" of People?

Jerry Bergman, Ph.D.

Genesis teaches that God pronounced the first two created people *very good* when He created them at the very beginning. "Then God said, 'Let us make man in our image, in our likeness.' So God created man in His own image, in the image of God He created him; male and female He created them. God blessed them and said to them, 'Be fruitful and increase in number; fill the earth and subdue it.' God saw all that He had made, and it was very good" (Genesis 1:26-31 NIV).

Soon after, Adam openly violated God's command not to eat of the forbidden fruit, and as a result, sin entered into the human race. God had to curse all of creation, and on that day Adam and Eve began the process of aging that always ends in death. As a result, an originally perfect created man began accumulating genetic mutations both in his body cells and in his germ cells.

Every generation has suffered from these mutations ever since. They degenerate each person's body, sometimes causing death through cancer and other diseases. Mutations in the germ line over many generations have caused degeneration of the entire human race. This process has continued until today. Geneticists have identified the mutations that cause over five thousand specific diseases in humans. Although a rare few mutations bring a benefit in very limited circumstances, 99.99% either cause harm or make virtually undetectable changes. But these small changes accumulate. After hundreds of generations, every person today inherits thousands of these mutations that now cause all kinds of damage.

Mutations in eggs and sperm cells are either lethal, harmful (disease-causing), or nearly neutral, having no immediate effect. As in body cells, near-neutral mutations cause miniscule damage. After enough of these accumulate, they cause a genetic meltdown leading to extinction of the species. The text *Principles of Medical Biochemistry*[184] under the subtitle "Mutations Are an Important Cause of Poor Health" states:

> At least one new mutation can be expected to occur in each round of cell division, even in cells with unimpaired DNA repair and in the absence of external mutagens [mutation-causing agents]. As a result, every child is born with an estimated 100 to 200 new mutations that were not present in the parents. Most of these mutations change only one or a few base pairs ... However, an estimated one or two new mutations are "mildly detrimental." This means they are not bad enough to cause a disease on their own, but they can impair physiological functions to some extent, and they can contribute to multifactorial diseases [when many causes add up to cause illness]. Finally, about 1 per 50 infants is born with a diagnosable genetic condition that can be attributed to a single major mutation (p. 153).

The authors concluded that, as a result:

> Children are, on average, a little sicker than their parents because they have new mutations on top of those inherited from the parents. This mutational load is kept in check by natural selection. In most traditional societies, almost half of all children used to die before they had a chance to reproduce. Investigators can only guess that those who died had, on average, more "mildly detrimental" mutations than those who survived (p. 153).

If macro-evolution is true, it is *going the wrong way!* It does not cause the ascent of life by adding new and useful biological coding instructions, but rather the descent of life by eroding what remains of the originally created biological codes. Should we call it "devolution" instead?

What do mutations have to do with "races?" Geneticists have studied DNA sequences in all kinds of different people groups. These studies reveal that each people group—which is most easily identified on a cultural level by sharing a specific language—shares a set of mutations. They must have inherited these "race" mutations from their ancestors after the Tower of Babel, since their ancestors freely interbred for the several hundred years between the Flood and the Tower. Amazingly, however, all these mutations make up less than one percent of all human DNA in the human genome. This means that no matter how different from you someone looks, they are 99.9% genetically identical to you. For this reason, even evolutionary geneticists admit that the term "race" has virtually no biological backing. It comes from cultural and mostly language differences. Bottom line: all peoples have the same genetic basis to be considered fully human, while expressing interesting cultural and subtle physical variations.

THE DNA BOTTLENECK

According to the chronologies in Genesis 5 and 11, the Genesis Flood occurred about 1,656 years after Creation. From possibly millions of pre-Flood peoples, only three couples survived the Flood and had children afterward. This caused a severe DNA bottleneck. Genetic bottlenecks occur when circumstances suddenly squeeze populations down to small numbers. They concentrate mutations and thus accelerate diseases. This occurs, for example, when people or animals marry or mate with close relations. Children or offspring from these unions have a much higher chance of inheriting mutations and the damage they cause. The genetic bottleneck of the Flood accelerated the decay of the human genome from Adam and Eve's once perfect genome.

Then, not long after the Tower of Babel, a major dispersion of humans occurred, leading to diverse ethnicities tied to languages. The Bible records about 70 families left the Tower. Many of them have gone extinct. Those few original languages have diversified into over 3,000 languages and dialects today. For example, English descended from the same basic language as German, while Welsh and Mandarin descended from fundamentally different original languages. Details from genetics and linguistics confirm Paul's statement in Acts 17:26, "He has made from one blood every nation of men to dwell on all the face of the earth."

Charles Darwin grouped these "nations" into "races," then organized races into those he believed were less human—less evolved—than others. He was completely wrong. Genetically, people in each ethnicity or nation share equal standing with other men. Biblically, they share equal standing before God, "For all have sinned and fallen short of the glory of God," according to Romans 3:23.

PHYSICAL DIFFERENCES

As noted, all of the differences between the human races are all very superficial, such as differences in skin, hair, and eye color. These traits account for less than 0.012% of human genetic differences, or 1 gene out of 12,000.[185] The two major racial differences that our society uses to label races are hair shape and skin color differences. One reason why we have two very distinct racial groups in America today, commonly called blacks and whites, is because the original immigrant population in the United States 350 years ago included primarily light-skinned people from Northern Europe and dark-skinned people from Africa. However, when dark-skinned people marry those with light skin, their children usually show medium-tone skin. Adam and Eve must have had medium tone skin. Sometime in history—probably at Babel—those with darker skin took their languages one direction, while those with lighter skin took theirs in another. Of course, they almost never remained in total isolation. Genetic tests reveal that probably everybody contains a mixture of ethnic-identifying mutations. Many dark Americans descended from dark-skinned African tribes that were kidnapped to be sold as slaves. Most people in the world have skin tones in between these two extremes, having brown skin and brown hair. Others have a mixture of traits.

HAIR

Subtle genetic differences develop different shaped hair follicles that produce from straight to curly human hairs. Round hair follicles manufacture tube-like, straight hair. Oval-shaped hair follicles produce flattened hair shafts, which curl. Flatter hairs make tighter curls.

SHAPE OF THE HAIR

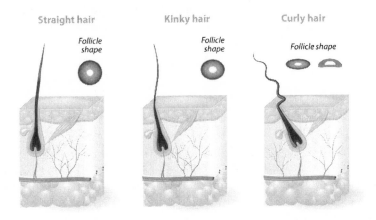

Figure 50. The Shape of the Hair (Credit: Dreamstime)

Human hair also shows a range of tones, from white to black, all depending on the amount of the pigment called melanin in hair. White hair, usually found in the elderly, almost totally lacks pigment. Brown hair contains a medium level, and black hair has the most amount of pigment. Red hair contains an iron oxide pigment which gives it the red-brown color similar to iron rust.

Special cellular machinery manufactures melanin pigments from the amino acid tyrosine. In humans, melanin serves mostly to add color to skin, hair, and eye irises. The chemical structure of melanin is rather complex, and so far it has defied detailed chemical analysis. In a similar way to how each snowflake differs from another, pigments like melanin are large enough to often include subtle molecular differences.

EYE COLOR

Melanin is responsible for the color of our eyes, which actually comes from the color that coats the iris diaphragm. The small black pupil of the eye is a hole that allows light to enter the inside of the eyeball, so it has no pigment. Light-sensitive photocells, called rods and cones, register light waves that enter the eyeballs. Variation in eye color from brown to green depends on the amount of melanin on the iris, which is determined genetically. However, it involves dozens of genes, each with its own inheritance pattern, so it is difficult to pinpoint the exact color of a child's eyes by the genes alone. Individuals with black or brown eyes have more melanin, which is important to block the sun's damaging ultraviolet rays. Blue eyes filter less ultraviolet light, which commonly damages retinas. Blue eyes are actually a result of a mutation that prevents adding the pigment necessary for proper eye protection. Persons with light blue, green, or hazel eyes have little protection from the sun, and so they often experience discomfort, irritation, burning, and tissue damage if the eyes are not protected by sunglasses when exposed to bright light. What does this have to do with ethnicities? First, eye color again illustrates how mutations cause damage. They are the biological enemies of human evolution. Second, the wide varieties and often stunning beauty in eye colors showcases God's creativity. Apes' and other animals' eyes are often simply dull in comparison.

SKIN COLOR

Like eye color, skin color depends on the level and type of melanin that special cells called melanocytes produce in the skin. In addition to showing variation, melanin protects the cell's nuclear DNA. It does not shield the entire cell, but it does cover the nucleus like a protective umbrella. Cells have molecular machines that detect and measure DNA damage caused by radiation. When excess damage occurs, they send their message to other systems that switch on melanin production. This causes skin to darken, or tan. No matter how dark one's skin normally is, if all the body systems work properly, skin will become darker after exposure to the sun's rays.

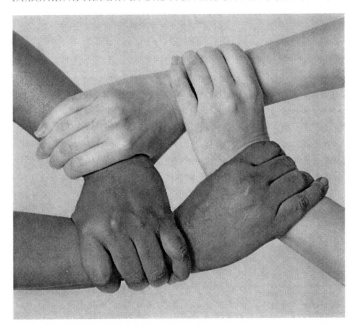

Figure 51. Skin Color tends to be a Major Factor in Determining Race
(Credit: Shutterstock)

Melanin reduces ultraviolet type B (UVB) damage by absorbing or scattering the ultraviolet radiation that otherwise would have been absorbed by the DNA, causing mutations. This protects against skin cancer. The specific wavelengths of light that melanin absorbs match those of DNA, thus protecting DNA from the sun's damaging radiation. Skin color also depends upon the size, number, shape, and distribution of melanocytes, as well as the chemical nature of their melanin content.

Modern genetics reveals that Adam and Eve could have had within their created genes almost all the pigmentation varieties seen today. If the trait of human skin color follows the "polygenic" inheritance pattern, then Adam and Eve's children could have appeared either very dark or very light, although most were probably medium brown, like their parents.

VITAMIN D TRIGGERED BY SUNLIGHT

A melanin balance is necessary to protect the skin's DNA from UV damage, yet allow the light skin to "trigger" its benefits. Skin harvests

UVB sunlight and uses it to process vitamin D, which the body requires. Vitamin D helps to promote proper bone density and growth by helping to regulate calcium and phosphorus in the body. Vitamin D deficiency leads to bones that lack the required calcium levels, causing rickets and even contributing to cancer, cardiovascular disease, mental impairment in older adults, and severe asthma in children.

What does all this have to do with the origin of people groups? As people migrated away from Babel in modern-day Iraq to northern latitudes, they had less exposure to sun. Others migrated to the tropics. Each person inherits their skin tone, and different skin tones interact differently with various climates.

Light-skinned people from the frozen north who visit lower latitude sunny locations have less melanin to block the sun's UVB rays. Without this protection, they may experience sunburn, which dramatically increases the odds of skin cancer. On the other hand, dark-skinned people visiting areas of dim sunlight may not produce enough vitamin D. They may need vitamin D supplements or obtain additional vitamin D from foods. For this reason, foods such as milk and bread are vitamin D fortified.

As global geographical distribution of various peoples shows, skin color variation is not determined by distance from the equator. Nevertheless, the skin tones we inherit can have different fits in different environments, and basic genetics reveal God could easily have programmed all human skin variation into the first created couple.

EYE SHAPE

Another example of superficial racial differences are the so-called almond eyes of Oriental people groups. The Asian eye has a fat layer in the upper eyelid that pushes the lid down, causing the eye to appear to be more closed. No Caucasian or Middle-Eastern ethnicities have this eye design, but two rare African tribes do. These tribes plus Asians must have inherited the trait from their ancestors at Babel. The information that codes for this trait was lost to Caucasians, Arabs and others who migrated away from those who retained it.

All of these are normal variations and examples of the remarkable variety that exists in all life—even within each created kind. Genetics confirm that only two people, Adam and Eve, contained all of the genes required to produce all of the variety seen across cultures today. In the end, as these people groups illustrate, race is not a biological, but a sociological construct.

DARWIN'S CONCLUSIONS ABOUT RACE AND SEX

Charles Darwin, the founder of modern evolutionary theory, openly expressed racist and gender sentiments that make modern readers cringe. As mentioned above, although the title of Darwin's most important book is often cited as *The Origin of Species*, the complete title is *The Origin of Species of Means of Natural Selection, or the Preservation of Favoured Races in the Struggle for Life*. The favored races, he argued in a later book titled *The Descent of Man* and *Selection in Relation to Sex*,[186] were supposedly Caucasians.

Darwin also taught that the "negro race" would become extinct, making the gap between whites and the lower apes wider. In his words:

> At some future period, not very distant as measured by centuries, the civilized races of man will almost certainly exterminate and replace throughout the world the savage races ... The break will then be rendered wider, for it will intervene between man in a more civilized state ... than the Caucasian, and some ape as low as a baboon, instead of as at present between the negro or Australian and the gorilla.[187]

Darwin did not begin racism, but his ideas bolstered it big time.[188] No science supports Darwin's ideas, and the Bible treats all people as equally human in God's sight.

Darwin also taught that women were biologically inferior to men and that human sexual differences were due, in part, to natural selection. As Darwin concluded in his *Descent of Man* book: "the average mental power in man must be above that of women." Darwin argued that the intellectual superiority of males is proved by the fact that men attain:

> a higher eminence, in whatever he takes up, than can women—whether requiring deep thought, reason, or

imagination, or merely the use of the senses and hands. If two lists were made of the most eminent men and women in poetry, painting, sculpture, music composition and performance, history, science, and philosophy, with half-a-dozen names under each subject, the two lists would not bear comparison ...We may also infer... that if men are capable of a decided preeminence over women in many subjects, the average of mental power in man must be above that of women.[189]

Modern society has proved this naïve assumption to be not only wrong, but also irresponsible. Darwin used many similar examples to illustrate the evolutionary forces that he concluded produced men to be of superior physical and intellectual strength and yet produce women to be more docile. Thus, due to "...success in the general struggle for life; and as in both cases the struggle will have been during maturity, the characters thus gained will have been transmitted more fully to the male than to the female offspring. Thus man has ultimately become superior to woman."[190] All this imaginative drivel ignores God's Word entirely. Genesis 1 extols the equality of genders by telling us that God created both husband and wife together as a married couple to reflect His image. It takes both to reflect His image. As a divinity student, Darwin surely read this. Did he deliberately ignore it?

CHAPTER 5

Are Humans and Chimps 98% Similar?

Jeffrey Tomkins, Ph.D. & Jerry Bergman Ph.D.

"As most people know, chimpanzees have about 98% of our DNA, but bananas share about 50%, and we are not 98% chimp or 50% banana, we are entirely human and unique in that respect."

–Professor Steve Jones, University College, London[191]

INTRODUCTION

One of the great trophies that evolutionists parade to prove human evolution from some common ape ancestor is the assertion that human and chimp DNA are 98 to 99% similar.[192] A quick Internet search reveals that this statistic is quoted in hundreds of textbooks, blogs, videos, and even scientific journals. Yet, any high school student can debunk the "Human and Chimp DNA is 98% similar" mantra that this chapter covers.

Why does this matter? Since we know that genes determine many aspects of our nature, from our sex to our hair color, then if we are genetically related to chimps, some may conclude that animal behavior by humans is expected, with no fear of divine judgment. But if we are all descended from Adam, not from animals, common animal behavior such as sexual promiscuity, cannot be justified on these grounds as some do.[193]

We will now review the major evidence that exposes the 98% myth and supports the current conclusion that the actual similarity is closer to

88%, or a difference of 12%, which translates to 360 million base pairs' difference. That is an enormous difference that produces an unbridgeable chasm between humans and chimpanzees.

If human and chimp DNA is nearly identical, why can't humans interbreed with chimps?[194] Furthermore, such an apparently minor difference in DNA (only 1%) does not account for the many obvious major differences between humans and chimps. Claiming that "because humans and chimps share similar DNA, they both descended from a recent common ancestor" is as logical as claiming that, "Watermelons, jellyfish, and snow cones consist of about 95% water, therefore they must have a recent common ancestor."

If humans and chimps are so similar, then why can't we interchange body parts with chimps? Over 30,000 organ transplants are made every year in the U.S. alone, and currently there are over 120,000 candidates on organ transplant lists—but *zero* of those transplants will be made using chimp organs!

Table 2. Organ Transplants[195]

Organ Transplants (2016)			
Organs	# Currently Waiting	% of Transplants Made Using	
		Human Organs	Chimp Organs
All Organs	121,520	100%	0%
Kidney	100,623	100%	0%
Liver	14,792	100%	0%
Pancreas	1,048	100%	0%
Kid./Panc.	1,953	100%	0%
Heart	4,167	100%	0%
Lung	1,495	100%	0%
Heart/Lung	47	100%	0%
Intestine	280	100%	0%

A BASIC OVERVIEW

The living populations of the chimp kind include four species that can interbreed. From the beginning, they were *soul-less* animals created on Day 6 of creation. Later that Day, God made a single man in His own image, and He gave him an eternal *soul* (Genesis 2:7). Then God commanded man to "rule over the fish in the sea and the birds in the sky, over the livestock and all the animals," including chimps (Genesis 1:26).

If the creation narrative from the Bible is true, we would expect *exactly* what we see in today's ape-kinds. First, all varieties of chimps have no concept of eternity. For example, they do not bury their dead nor do they conduct funeral rituals. Secondly, apes use very limited verbal communication—they cannot write articles or even sentences. Thirdly, they do not display *spiritual or religious practices* as humans do. In other words, they show no capacity for knowing their spiritual creator through worship or prayer. This fits the Biblical creation account that humans are created, spiritual beings with a soul.

It is logical that God, in His desire to create diverse life forms on Earth, would begin with the same building materials, such as DNA, carbohydrates, fats, and protein, when making various animal kinds. Research has revealed that He used similar building blocks for all of the various physical life forms that He created. Genetic information in all living creatures is encoded as a sequence of only 4 nucleotides (guanine, adenine, thymine, and cytosine, shown by the letters G, A, T, and C). We also see this principle in nature—such as many plants sharing Fibonacci spirals (clear numerical patterns) and sequences as basic building blocks and patterns.

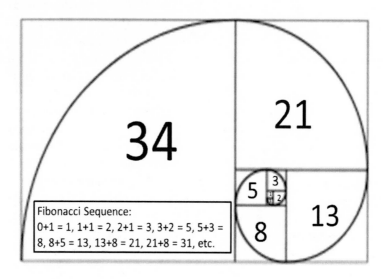

Figure 52. Fibonacci Number Sequence. A Fibonacci spiral approximates the golden spiral using quarter-circle arcs inscribed in squares of integer Fibonacci-number side, shown for square sizes 1, 1, 2, 3, 5, 8, 13, 21, 34 etc.

Figure 53. Examples of the Fibonacci Sequence in Nature (Credit: Wikipedia)

Chimp and human DNA share many similarities, but this does not prove that those similarities came from shared ancestors. **They are similar due to design constraints that require an engineer to use many of the same raw materials and building plans to produce two very different types of "machines."** For example, an automotive engineer could make a Volkswagen bug and a Porsche Carrera framework out of steel, glass, and plastic but not diatomic oxygen, carbon dioxide, and H_2SO_4. Next, let's take a look at just how different chimps and humans are, even though they share some similar DNA.

COMPARISONS OF CHIMPS AND HUMANS

A child that sees a chimpanzee can immediately tell that it is radically different from a human. Compared to chimps, humans are about 38% taller, are 80% heavier, live 50% longer, and have brains that are about 400% larger (1330 ccs compared to 330 ccs).[196] Look at someone next to you and roll your eyes at them. Chimps can't do that because their sclera, like most other animals, is hidden behind their eyelids. Now tap your fingertips with your thumb. Chimps can't do that either—their fingers are curved, their thumbs are both tiny and set further back on their wrists than humans, and they are missing the flexor pollicis longus—the major muscle that controls thumb dexterity in humans. Additionally, their knees point out, whereas ours point forward. Humans can build space shuttles and write songs. Chimps cannot.

Scientists now know that chimpanzees are radically different than humans in many different ways besides their outward appearance. Humans and chimpanzees have differences in bone structures, in brain types, and in other major parts of their physiology. Humans also have the ability to express their thoughts abstractly in speech, writing, and music, as well as developing other complicated systems of expression and communication. This is why humans stand above all other types of creatures.

The claimed small genetic differences between human and chimp DNA (1 to 2%) must account for these and many other major differences! The difference between humans and chimpanzees is major and includes about 350 million different DNA bases. **In fact, it is hard to compare the two genomes because they are so different.**

The chimp genome is much longer than the human genome: The chimp genome is not 98% of the length of the human genome. According to the latest data, there are 3,096,649,726 base pairs in the human genome and 3,309,577,922 base pairs in the chimpanzee genome (a 6.4% difference).[197] Telomeres in Chimps and other apes are about 23 kilobases (a kilobase is 1,000 base pairs of DNA) long. Humans are unique among primates with much shorter telomeres only 10 kilobases long.[198] The human Y chromosome is a very different size and has many markers that do not line up when the human and chimpanzee chromosome is compared.[199] Even if human and chimpanzee DNA sequences are as similar as some evolutionists claim, the DNA coding makes two entirely different creatures!

Humans have forty-six chromosomes, while chimps have forty-eight. Additionally, the fusion theory (i.e., the claim that human chromosome 2 was created by the fusion of two smaller chimpanzee chromosomes) has now been refuted. In fact, this claim has been used as veritable proof of common ancestry by many.

Research by Dr. David A. DeWitt has revealed new stunning insights regarding the major differences between human and chimp DNA: There exist 40–45 million bases [DNA "letters"] in humans missing from chimps, and about the same number present in chimps that are absent from man. These extra DNA nucleotides are termed "insertions" and "deletions" because they are assumed to have been added or lost from the original common ancestor sequence. These differences alone put the total number of DNA differences at about 125 million. However, since the insertions can be more than one nucleotide long, about 40 million total separate mutation events would be required to separate the two species. Such research continues to reveal that we are genetically far more different from chimps than the textbooks reveal! To put this number into perspective, a typical 8½ x 11-inch page of text has about 4,000 letters and spaces. It would require 10,000 such pages of text to equal 40 million letters! These "10,000 pages" of different DNA programming are also enough to fill 20 full-sized novels.

The difference between humans and chimpanzees includes about 45 million base pairs in the human that are absent from the chimp, and about 45 million base pairs in the chimp absent from the human.[200] More research has left no doubt that a specific set of genetic programming exists for humans, and another specific set exists for chimps. Despite clear differences between humans and apes, we are repeatedly told by an array of mainstream outlets including textbooks, that human and chimpanzee DNA is 98 to 99% similar. Are we really just a few genetic changes away from being an ape? And what is the field of modern genetics research actually revealing?

Biology textbooks typically explain that humans descended from some common ancestor related to the great apes. This animal group consists of orangutans, gorillas, and chimpanzees. Of these apes, evolutionists claim that humans are most closely related to chimpanzees based on comparisons of human DNA to chimp DNA. The real world consequences of this ideology involve concluding that humans are not special creations, but that they are evolved animals.

This has been a primary foundation for the mistreatment of humans worldwide by genocidal political leaders and governments over the past 150 or so years. One highly reputable study showed that the leading cause of death in the 20th century was "Democide"—or "murder by government," which has claimed well over 260 million lives.[201] All of the totalitarian murderous tyrannies the world over, despite their different political variations, maintained the same Darwinian evolutionary philosophy that humans are higher animals to be herded and culled in wars, death-camps, abortions, mass starvations, and outright slaughter.[202]

Do the new sciences of DNA sequencing and genomics justify the evil ideology that comes from believing that some humans are more evolved while others are nothing but common animals? Genetics research exposes human evolution as a total misrepresentation of reality. If this question is important to you—and it should be as a member of the human family—you will find this section very important. Once you understand that the new DNA evidence debunks the alleged human evolution paradigm, you will appreciate that you are a unique creation whom the Creator made in His own image. You are special and unique compared to all of creation.

When experts talk about DNA similarity, they refer to a variety of different features. Sometimes they talk about humans and chimpanzees having the same genes. At other times, they talk about certain DNA sequences being 98 to 99% similar. First, let's consider why human and chimpanzee DNA sequences are actually closer to 88% than 98% similar. Then, describing the concepts of genes and gene similarity will reveal much insight into human and chimp DNA dissimilarity.

REALITY OF DNA AND GENOME SIMILARITY

Human, plant, and animal DNA is packaged into separate packages called chromosomes. Each one contains millions of the four different DNA bases (T, A, C, G), stacked like rungs on a ladder. Their specific order forms a complex set of instructions called the "genetic code." Humans have two copies of each chromosome; one set of 23 from the mother and one set of 23 from the father. Each chromosome set contains over 3 billion base pairs of information. Therefore, a total of 6 billion DNA bases are in our 46 chromosomes that are inside of nearly every cell in our body. When scientists talk about a creature's genome, they are only referring to one set of chromosomes. Thus, the reference genome in humans is the sum total of one complete set of 23 chromosomes.

The "initial draft" of DNA sequences in the human genome was initially published in 2001. In 2004, scientists published a more complete version, but there were still small parts that remained to be sequenced, so researchers kept updating the human genome as DNA sequencing technologies improved and more data were acquired. The human genome is now one of the most complete of all known genome sequences—mostly because considerably more research money has been spent on it compared to other life forms.

In order to organize 3 billion bases, researchers use unique DNA sequences as reference markers. Then they determine where these short sequences are located on each chromosome. They assumed that comparing sequences between related creatures would help locate them. Scientists initially chose chimpanzees as the closest creature to humans because they knew that their proteins and DNA fragments had similar biochemical properties.[203] However, some researchers for various reasons chose gorillas or orangutans as being closest to humans and compared their DNA instead. In fact, a recent research paper made the claim that orangutans' DNAs were more similar to humans' DNA in structure and appearance than chimpanzee, and thus orangutans should be considered our closest ancestor. Nevertheless, the consensus among evolutionary scientists is that chimpanzees are closest to humans on the hypothetical evolutionary tree. For this reason, most genetics studies assume this relationship before they even begin analyzing DNA.

In the early days of DNA sequencing, in the 1970s, scientists were able to sequence only very short segments of DNA. For this reason, they focused on DNA segments that they knew would be highly similar between animals, such as blood globin proteins and mitochondrial DNA (DNA which is inherited from the mother). They selected similar regions for comparison, because you cannot glean any meaningful comparisons between two DNA sequences that exist only in one and not the other. Researchers discovered that many of the short stretches of DNA genetic sequences that code for common proteins were not only highly similar in many types of animals, but that they were nearly identical between humans and apes.[204]

Before the true levels of similarity between human and chimp genomes can be determined, a basic understanding of what DNA sequencing actually entails is helpful. While the basic DNA sequencing techniques have not changed much since they were developed, the use of small-scale robotics and automation now enable researchers to sequence massive amounts of small DNA fragments. The DNA of an entire organism is too long to be sequenced, thus millions of small pieces, hundreds of bases in length are sequenced. Computers are then used to assemble the small individual pieces into larger fragments based on overlapping sections.[205] DNA regions that have hundreds of repeating sequences are, for this reason, very difficult to reconstruct, yet we now know that they are important for cell function.

ENTER NEW TECHNOLOGY

Despite the early discoveries of apparently high DNA similarity between humans and chimps, large-scale DNA sequencing projects began to present a very different picture. In 2002, a DNA sequencing lab produced over 3 million bases of chimp DNA sequence in small 50 to 900 base fragments that were obtained randomly from the entire chimp genome.[206] The short sequences must then be assembled, and the physical arrangement of chimp DNA sequences are largely based on the human genomic framework.[207] This turned out to be only one of many problems. When the chimp DNA sequences were matched with the human genome by computers, only two-thirds of the DNA sequences could be lined up with human DNA. While many short stretches of DNA existed that were very similar to human DNA, more than 30% of the chimp DNA sequence was not similar to human DNA!

In 2005, the first rough draft of the chimpanzee genome was completed by a collaboration of different labs.[208] As a rough draft, even after the computational assembly based on the human genome, it still consisted of thousands of small chunks of DNA sequences. The researchers then assembled all of the small sequences of chimp DNA together to form a complete genome. They did this by assuming that humans evolved from a chimp-like ancestor, so they used the human genome as the framework to assemble the chimp DNA sequences.[209] At least one lab that helped to assemble the chimp sequence admitted that they inserted human DNA sequences into the chimp genome based on the evolutionary assumptions. They assumed that many human-like sequences were missing from the chimp DNA, so they added them electronically. The published chimp genome is, thus, partly based on the human genome. Because it contains human sequences, it appears more human than the chimp genome in fact is.

A large 2013 research project sequenced the genomes of chimpanzees, gorillas, and orangutans to determine their genetic variation. Then they assembled all of these genomes using the human genome as a framework![210] Much shorter lengths of DNA fragments are produced by new technologies, providing faster results, but those smaller sections are more difficult to assemble.

Unfortunately, the research paper describing the 2005 chimp draft genome avoided the problem of overall average genome similarity with humans by analyzing the regions of the genomes that were already known to be highly similar. This deceptively reinforced the mythical 98% similarity notion. However, enough data were in the 2005 report to allow several independent researchers to calculate overall human-chimp genome similarities. They came up with estimates of 70 to 80% DNA sequence similarity.[211]

This result is important because evolution has a difficult time explaining how only 2% of 3 billion bases could have evolved in the 6 million years since they believe chimps and humans shared a common ancestor. They want to avoid the task of explaining how 20 to 30% of three billion bases evolved in such a short time! Natural processes cannot create 369 million letters of precisely coded information in a billion years, let alone a few million years.[212]

Thus, the commonly reported high levels of human-chimp DNA similarity were actually based on highly similar regions shared by both humans and chimps and exclude vastly different regions of these separately created genomes. Cherry-picking of data is not valid science. Other published research studies completed between 2002 and 2006 compared certain isolated regions of the chimp genome to human DNA. These also seemed to add support to the evolutionary paradigm, but reinserting dissimilar DNA sequence data where it could be determined that evolutionists had omitted it from their analyses, significantly changed the results.[213] The results showed that the actual DNA similarities for the analyzed regions varied between about 66% to 86%.

One of the main problems with comparing DNA segments between different organisms that contain regions of strong dissimilarity is that the computer program commonly used (called BLASTN) stops matching DNA when it hits regions that are markedly different. These unmatched sections consequently are not included in the final results, raising significantly the overall similarity between human and chimp DNA.

In addition, the computer settings can be changed to reject DNA sequences that are not similar enough for the research needs. The common default setting used by most evolutionary researchers kicks out anything less than 95% to 98% in similarity. In 2011, Tompkins compared 40,000 chimp DNA sequences that were about 740 bases long and already known to be highly similar to human.[214] The longest matches showed a DNA similarity of only 86%.

If chimp DNA is so dissimilar to human, and the computer software stops matching after only a few hundred bases, how can we find the actual similarity of the human and chimp genomes? A 2013 study resolved this problem by digitally slicing up chimp DNA into the small fragments that the software's algorithm could optimally match.[215] Using a powerful computer dedicated to this massive computation, all 24 chimp chromosomes were compared to humans' 23 chromosomes. The results showed that, depending on the chromosome, the chimp chromosomes were between 43% and 78% similar to humans. Overall, the chimp genome was only about 70%[216] similar to human. These data confirmed results published in secular evolutionary journals, but not popularized by the media or evolutionists.

Although textbooks still contain the 98% DNA similarity claim, scientists in the human-chimp research community now recognize the 96% to 98% similarity is derived from isolated areas. However, while the 96–98% similarity is crumbling, geneticists rarely make public statements about overall estimates because they know it would debunk human evolution. Although the human and chimpanzee genomes overall are only about 88% similar, some regions have high similarity, mostly due to protein-coding genes. Even these high similarity areas actually have only about 86% of matching sequences overall when the algorithm used to analyze them is set to produce a very long sequence match.[217]

The regions of high similarity can be explained by the fact that common genetic code elements are often found between different organisms because they code for genes that produce proteins with similar functions. For the same reason that different kinds of craftsmen all use hammers to drive or pry nails, different kinds of creatures use many of the same biochemical tools to perform common cellular functions. The genome is a very complex system of genetic codes, many of which are repeated in organisms with similar functions. This concept is easier to explain to computer programmers and engineers than biologists who are steeped in the evolutionary worldview.

GENE SIMILARITIES—THE BIG PICTURE

If two creatures have the same genes, usually only a certain part of a gene sequence is shared. The entire gene could be only 88% similar, while a small part of it may be 98% similar. In fact, the protein-coding regions called "exons" are on average in humans only about 86% to 87% similar to chimps. Much of this is due to human exon sequences completely missing in chimps.

The original definition of a gene describes it as a DNA section that produces a messenger RNA that codes for a protein. Early estimates projected that humans contained about 22,000 of these protein-coding genes, and the most recent estimates are about 28,000 to 30,000.[218] We now know that each of these protein-coding genes can produce many different individual messenger RNA variants due to gene regulation of gene section splicing variations. Consequently, over a million protein varieties can be made from 30,000 or fewer genes! Nevertheless, less than 5% of the human genome contains actual "exon" protein-coding sequences.

HUMANS HAVE A HIGH LEVEL OF DNA/ GENE SIMILARITY WITH MULTIPLE OTHER CREATURES

The human body has many molecular similarities with other living things because they are all made up of the same molecules, they all use the same water and atmosphere, and they all consume foods consisting of the same molecules. Their metabolism, and therefore their genetic make-up, would resemble one another. However, this is not evidence that they evolved from a common ancestor any more than all building constructed using common materials (brick, iron, cement, glass, etc.) indicates that the buildings share a common ancestor.

The same holds for living beings. DNA contains much of the information necessary for the development of an organism. If two organisms look similar, we would expect there to be some DNA similarity. The DNA of a cow and a whale should be more alike than the DNA of a cow and a bacterium. Likewise, humans and apes have many morphological similarities, so we would expect there would be many DNA similarities. Of all known animals, chimps are most like humans, so we would expect that their DNA would be most like human DNA.[219]

This is not always the case, though. Some comparisons between human DNA/genes and other animals in the literature including cats have 90% of homologous genes with humans, dogs 82%, cows 80%,[220] chimpanzees 79%, rats 69%, and mice 67%.[221] Other comparisons found include fruit fly (Drosophila) with about 60%[222] and chickens with about 60% of genes corresponding to a similar human gene.[223] One should keep in mind that these estimates suffer from the same problems that the estimates comparing humans to chimps do.

THE MYTH OF "JUNK" DNA

The 30,000 or so genes of the human genome occupy less than 5% of the 3 billion total base pairs in the human genome. Because evolutionary scientists did not know what the other 95% of the genome does and because they needed raw genetic material for evolution to tinker with over millions of years, they labeled it "junk DNA." However, new research from different labs all over the world has documented that over 90% of the entire human genome is transcribed into a dizzying array of RNA molecules that perform many different important functions in the cell.[224] This phenomenon, called "pervasive transcription," was discovered in an offshoot of the human genome project called ENCODE, which stands for ENCyclopedia of DNA Elements.[225]

While refuting "junk" DNA, the ENCODE project has also completely redefined our concept of a gene. At the time of this writing, experts estimate that non-protein-coding RNA genes called *long noncoding RNAs* or "lncRNAs" outnumber protein coding genes at least 2 to 1.[226] They have similar DNA structures and control features as protein-coding genes, but instead they produce functional RNA molecules that do many things in the cell.

Some regulate the function of protein coding genes in various ways and remain in the cell nucleus with the DNA. Others are transported into the cell cytoplasm to help regulate various cellular processes in collaboration with proteins. The cell exports other lncRNAs outside of the cell in which they are produced. There they regulate other cells. Many of these lncRNA genes play important roles in a process called epigenetics, which helped to regulate many aspects of how chromosomes are organized and the genome functions.

In contrast to many evolutionary studies that compared only the highly similar protein-coding regions of the genome, the lncRNA regions are only about 67 to 76% similar—about 10 to 20% less identical than the protein-coding regions. Chimp and human lncRNAs are very different from each other, but they are critical to each life form kind.

Clearly, the *entire genome* is a storehouse of important information.

Using the construction project analogy, the protein-coding genes are like building blocks, and the noncoding regions regulate and determine how and where the building blocks are used. This is why the protein-coding regions tend to show more similarities between organisms and the noncoding regions show fewer similarities. Protein-coding regions specify skin, hair, hearts, and brains, but "noncoding" regions help organize these components into the different but distinct arrangements that define each creature's body plan. Given all these facts, it is not surprising that humans and chimps are markedly different!

CHROMOSOME FUSION DEBUNKED

One of the main arguments that evolutionists use to support their human-chimp story is the supposed fusion of two ape-like chromosomes to form human chromosome number two. The great apes actually contain two more (diploid) chromosomes than humans—humans have 46 and apes have 48. Large portions of two small ape chromosomes look somewhat similar to human chromosome 2 when observed under a microscope after special staining. Evolutionists attempt to argue that they look so similar because they have descended from a common ancestor, namely two ancient chromosomes from an ape-like ancestor fused during human evolution.[227]

Supposedly, the modern chimp's chromosomes look like the imaginary ape-human ancestors' did. Taking their cues from evolutionary assumptions, these two chimp chromosomes are called 2A and 2B. Gorillas and orangutans also have a 2A and 2B chromosome like chimps. Could the similarities between these two ape chromosomes and human chromosome 2 come from some cause other than common ancestry? What detailed features would we expect to see if these chromosomes fused to become one in humans?

In 1991, scientists found a short segment of DNA on human chromosome 2 that they claimed was evidence for fusion. It looked to them like a genetic scar left over from two chromosome ends that were supposedly stitched together, even though it was not what they should have expected based on the analysis of known fusions in living mammals.[228] The alleged fusion sequence consisted of what looked like a degraded head-to-head fusion of chromosome ends called "telomeres."

Telomeres contain repeats of the DNA sequence TTAGGG over and over for thousands of bases. Human telomeres are typically 5,000 to 15,000 bases long. If these actually fused, then they should have thousands of TTAGGG bases.[229] The alleged fusion site, however, is only about 800 bases long and only 70% similar to what would be expected. Plus, telomeres are specifically designed to prevent chromosomal fusion, and this is why a telomere-telomere fusion never has been observed in nature!

This fusion idea has for many years been masquerading as a solid argument proving human evolution from a chimp-like ancestor, but it has now been completely refuted by genetic research. It turns out the alleged fusion site is actually a *functional* DNA sequence inside an important noncoding RNA gene.[230] These DNA sequences are termed "noncoding" sequences because they do not code for genes that produce proteins. However, these sequences often code for useful and often critically important RNAs.

In 2002, researchers sequenced over 614,000 bases of DNA surrounding the supposed fusion site and found that it was in a gene-rich region. Also, the fusion site itself was inside of what they originally labeled a pseudogene, which are supposedly damaged "dysfunctional relatives" of formerly real protein-coding genes.[231] They are supposed to represent more genetic junk from a messy evolutionary past. However, continual discoveries of important cellular roles for "pseudogenes" keep surprising evolutionists, who expect junk, but keep finding functional genetic design.

New research using data from the ENCODE project now shows that part of the so-called "fusion site" is part of a noncoding RNA gene that is expressed in many different types of human cells. The research also shows that the alleged fusion site encodes a location inside the gene that binds to proteins that regulate the gene expression. Even more clear evidence for creation is the finding that not one of the other genes within 614,000 bases surrounding the alleged fusion site exists in chimpanzees. Although many evolutionists, unaware of the recent research, still promote it, the facts reveal that human chromosome 2 was a unique creation showing none of the expected signs of a chromosome fusion.

BETA-GLOBIN PSEUDOGENE DEBUNKED

Another story that evolutionists use to promote human-ape ancestry is the idea of shared mistakes in supposedly broken genes called pseudogenes noted above. Supposedly, the ape ancestor's genes were first mutated. Then, after its descendants diverged, both its chimp and human descendant genomes have retained those old mutations. After all, they argue, how else could two different but similar species have the same mutations in the same genes unless they evolved from the same ancestor?

If this story was true, if we evolved from apes, then we were obviously not created in God's image. Fortunately, exciting new research shows why science supports Scripture. As noted, many so-called "pseudogenes" are actually very *functional*. They produce important noncoding RNAs discussed previously.[232] This means that the shared DNA sequence "mistakes" were actually purposefully created DNA sequences all along.

One example is the beta-globin pseudogene, actually a functional gene in the middle of a cluster of five other genes. The other five genes code for and produce functional proteins. Evolutionists originally claimed that the beta-globin gene was broken because it did not produce a protein and because of its DNA similarity to chimps and other apes. Now multiple studies have shown that it produces long noncoding RNAs and is the most genetically networked gene in the entire beta-globin gene cluster, meaning it is transcribed often, likely for multiple purposes.[233]

Similar to the way that computer servers are connected to each other to produce the internet, genes do not act alone, but they are functionally connected to many other genes in the genome. Not only do other genes depend on the proper function of the beta-globin pseudogene, but over 250 different types of human cells actively use the gene! Why do chimps and humans share this very similar sequence? Not because they both inherited it from a common ancestor, but because they both use it for very similar purposes, like bricks can be used to build either a house or a library.

GULO PSEUDOGENE DEBUNKED

Another case of so-called evidence for evolution is the GULO pseudogene, which actually looks like a broken gene. A functional GULO gene produces an enzyme in animals that helps to make vitamin C. Evolutionists claim that humans, chimps and other apes share GULO genes that mutated in the same places because the mutations occurred in their common ancestor.

However, broken GULO pseudogenes are also found in mice, rats, bats, birds, pigs, and famously, guinea pigs. Did we evolve from guinea pigs? When the GULO gene was recently analyzed in its entirety, researchers found no pattern of common ancestry.[234] Instead, it looks like this gene is predisposed to being mutated no matter what creature it is in. Since humans and other animals can get vitamin C from their diet, they can survive without the gene. Also, the other genes in the GULO biochemical pathway produce proteins that are involved in other important cellular processes. Losing them could be disastrous to the organism. So many creatures and humans can tolerate having a damaged GULO gene by consuming plenty of vegetables with vitamin C.

The GULO gene region and the mutational events that damaged it are associated with unique categories of a system that use transposable elements. These are commonly called "jumping genes," and they can cut themselves out of one location in the genome and splice themselves into another location. The many different types of transposable elements in the human genome serve very important tasks. Sometimes, though, they splice themselves into the wrong location and disrupt genes.

In the case of GULO, the transposable element patterns between humans and each of the ape kinds that were evaluated show unique differences. Therefore, GULO shows no pattern of common ancestry for humans and apes—negating this evolutionary argument. Like the claims of 99% similarity, chromosome fusion, and Beta-globin, evolutionists built the GULO argument based on a prior belief in evolution, plus a lack of knowledge about how these systems actually function in cells.

In reality, the GULO pseudogene data defies evolution and vindicates the creation model. According to the Genesis account of the fall that caused the curse on creation, we would expect genes to mutate as this one did. This process of genetic decay, called genetic entropy, is found everywhere in the animal kingdom. Cornell University Geneticist John Sanford has shown in several studies that the human genome shows no signs of evolving or getting better, but is actually in a state of irreversible degeneration.[235] Perhaps our early ancestors had a working GULO gene that could thus manufacture vitamin C. Today, lacking sufficient vitamin C in our diets causes an illness called "scurvy."

THE HUMAN-CHIMP EVOLUTION MAGIC ACT

Stage magicians, otherwise known as illusionists, practice their trade by getting you to focus on some aspect of the magician's act to divert your focus from what is really occurring or what the other hand is doing. By doing this, they get you to believe something that isn't true, creating an illusion—a fake reality. The human-chimp DNA similarity "research" works almost the same way.

The evolutionist who promotes the human-chimp fake paradigm

of DNA similarity accomplishes the magic act by getting you to focus on a small set of data representing bits and pieces of hand-picked evidence. In this way, you don't see the mountains of hard data that utterly defy evolution. While some parts of the human and chimpanzee genomes are very similar—those that the evolutionists focus on—the genomes overall are vastly different, and the hard scientific evidence now proves it. The magic act isn't working any longer, and more and more open-minded scientists are beginning to realize it.

CONFRONTING HUMAN-CHIMP PROPAGANDA

To close this chapter, let's discuss a hypothetical exchange that could take place using the information in this chapter with some human-chimp similarity proponent. This exchange could happen with a teacher, a friend, or a schoolmate. First, the person makes the claim that "human and chimp DNA is genetically 98–99% identical or similar." You can respond, "That's only partially true for the highly similar regions that have been compared between humans and chimps." You can then clarify this response by noting that "recent research has shown that, overall, the entire genome is only about 88% similar on average when you include all the DNA. This is equal to 12 percent difference, or 360 million base pair differences."

You can also add, "Several thousand genes unique to humans are completely missing in chimps, and scientists have found many genes that are unique to chimps are missing in humans." Then ask, "How can you explain these massive differences by evolutionary processes?" In sum, ask, "How is it that such supposedly minor differences in DNA can account for such major and obvious differences between humans and chimps?"

At this point in the conversation, you will rapidly find out if the person is really interested in learning more about the issue of human origins, or if they are so zealous about evolutionary beliefs that they refuse to listen to challenging evidence. In reality, the whole modern research field of genetics and genomics is the worst enemy of evolution. As new genomes of different kinds of organisms are being sequenced, they consistently are shown to be unique sets of DNA containing many genes and other sequences that are specific to that type of creature. Evolutionists call these new creature-specific genes "orphan genes" because they are not found in any other type of known creature.[236] Orphan genes appear suddenly in the pattern of life as unique sections of genetic code with no evolutionary history. Of course, believers in an omnipotent Creator know that each different genome, such as that for humans and that of chimpanzees, was separately, uniquely, and masterfully engineered at the beginning of creation. God created and embedded each creature's orphan genes to network with all the rest of that creature's genetic coding instructions. The scientific data overwhelmingly show that God deserves the credit, and evolution deserves none.

CONCLUSION

With so much at stake, like the answer to life's largest question, "Where did I come from?", do we want to trust in extremely biased answers? Every high school student can refute 98% similarity dogma by tracking the main points above as outlined below.

1. Overall, the entire genome is only about 88% similar on average when you include all the DNA. This is equal to a 12% difference, or 360 million base pair differences, and it is a chasm away from our supposed closest evolutionary relative.
2. The "Junk" DNA claim has long been refuted and most of it has been found to have clear functions which are regulatory in nature.
3. The Chromosome Fusion claim is false. First, telomeres are designed not to fuse, thus, as per their design, telomere to telomere fusion is unknown in the natural world. Telomeres contain repeats of the DNA sequence TTAGGG over and over for thousands of bases. Human telomeres are from 5,000 to 15,000 bases long. If these actually fused, then they should have thousands of TTAGGG bases, but the alleged fusion site is only about 800 bases long and only 70% similar to what would be expected. The claimed fusion site actually contains a gene, and that gene is very different from a telomere.
4. The Beta-globin Pseudogene is not a pseudogene and not proof that a damaged gene was inherited from the human-chimp common ancestor. It is actually a functional gene in the middle of a cluster of five other genes.
5. The GULO Pseudogene is not evidence for common decent, but evidently it is due to a hot spot in this gene, meaning that it is in an area of the genome that is very prone to mutate.

CHAPTER 6

How Modern Genetics Supports a Recent Creation

Jeffrey P. Tomkins, Ph.D.

A substantial amount of convincing evidence exists for a recent creation as described in the Bible using hard scientific data from the disciplines of geology, paleontology, physics, and astronomy.[237] However, what does the field of genetics and modern genomics, one of the most rapidly advancing areas of science have to offer in this regard? As it turns out, new discoveries using the tools of modern biotechnology also showcase recent creation and events associated with the global flood.

One of the first questions we need to ask ourselves in order to form a hypothesis or model about the origins of human genetics involves comparing the *predictions* of creation science versus evolution. Creation science predicts that the genomes of all of the different kinds of living creatures were created perfect *in the beginning*. However, due to the curse on creation related to man's sin and rebellion combined with the damaging effects of time, we should see degradation, corruption, and the loss of information. While evolutionists do recognize that information loss occurs, in the overall grand Darwinian scheme, their model predicts just the opposite of creationists. They believe that over vast amounts of time, genomes evolved and became more complex—gaining new information through random mutational processes. Let's see what the data actually says and which prediction or model is supported by it.

Over time, errors are made by cellular machinery that copies DNA during the standard process of cell division. These errors are called mutations. Sometimes they can lead to serious diseases such as cancer. However, when a mutation occurs in cell division that leads to making sperm or egg cells, these mutations can be inherited and passed on to the next generation. In fact, scientists have actually measured this rate among humans and found it to be about 75 to 175 mutations per generation.[238] An earlier chapter in this book quoted an estimate of between 100 and 200 new mutations. Using this known data about mutation rates, a research group lead by Cornell University geneticist Dr. John Sanford modeled the accumulation of mutations in the human genome over time using computer simulations that accurately accounted for real-life factors. They incorporate the standard observations and theories behind population genetics. They found that the buildup of mutations would eventually reach a critical level and become so severe that humans would eventually go extinct. This process of genome degradation over time and successive generations is called *genetic entropy*. Remarkably, the timeframe of human genome degradation coincides closely with a recent creation of six to ten thousand years ago as predicted by the documented genealogies found in the Bible.[239]

Amazingly, after the results of the human genome modeling research were published, two different large groups of scientists unwittingly vindicated the idea of genetic entropy and a recent creation.[240] In each study, they sequenced the protein coding regions of the human genome. One study examined 2,440 individuals and the other 6,515. From the DNA sequence data, they discovered many single nucleotide differences (variants) between people in their protein coding genes, with most of these being very rare types of variants. In addition, they found that over 80% of these variants were either deleterious or harmful mutations. Surprisingly, they attributed the unexpected presence of these harmful mutations to "weak purifying selection." This essentially means that the alleged ability of natural selection to remove these harmful variants from human populations was powerless to do so. Sanford's model predicted that natural selection could not remove these slightly harmful mutations, and these studies confirmed exactly that in the real world of human genetics.

Not only were these studies bad news for the evolutionary idea of mutation and natural selection as the supposed drivers of evolutionary change, but they also overwhelmingly illustrated genetic entropy. Most of the mutations resulted in heritable diseases afflicting important protein-coding genes. Protein-coding regions are less tolerant of variability than other parts of the genome. These DNA regions can give us a better idea of the gene sequences of our forefathers because those who have too many mutations to important genes died long ago. Secular evolution-believing scientists usually pin their models of DNA change over time, referred to as molecular clocks, to millions of years before they even approach the data. In other words, they assume millions of years of human evolution and literally incorporate these deep time numbers into their models. The millions of years conclusion does not come from biology experiments.

In contrast, these new genetic variant studies used models of human populations that incorporate more realistic data over known historical time and geographical space. The resulting data revealed a very recent, massive burst of human genetic diversification. Most of it links with genetic entropy. One of the research papers stated, "The maximum likelihood time for accelerated growth was 5,115 years ago."[241] This places the beginning of the period of genetic diversification of humans close to the Genesis Flood and subsequent dispersion at the Tower of Babel, a point in time that the earth began to be repopulated through Noah's descendants. This recent explosion of human genetic variability is clearly associated with genetic entropy, and it also follows the same pattern of human life expectancy that rapidly declined after the Flood as also recorded in the Bible.[242]

One more important realm of research demonstrating a recent creation comes from Harvard trained scientist Dr. Nathaniel Jeanson. He has been examining the mutation rates of DNA in mitochondrial genomes.[243] The mitochondria are located outside the cell's nucleus. Mitochondria provide the energy for cells. They also contain their own DNA molecule that encodes a variety of proteins it uses for energy processing. The mitochondrial DNA molecule is typically inherited from the egg cell from a creature's mother. Its mutation rates can accurately be measured to produce a molecular-genetic clock. When these genetic clocks are not calibrated by (theoretical) evolutionary timescales, but by using the organism's observed mutation rate, we can reveal a more realistic and unbiased estimate of that creature's genetic life history. By comparing the molecular clock rates in a few very different animals (fruit flies, round worms, water fleas, and humans), Dr. Jeanson demonstrated that a creation event for these organisms (including humans) occurred not more than 10,000 years ago!

Interestingly, buried deep within a secular research paper in 1997, the same thing regarding human mitochondrial DNA (mtDNA) mutation rate was reported, but it received little attention in the media. The authors of this paper wrote, "Using our empirical rate to calibrate the mtDNA molecular clock would result in an age of the mtDNA MRCA [the first human woman] of only ~6,500 years…"[244] One year later, another author wrote in the leading magazine *Science*, "Regardless of the cause, evolutionists are most concerned about the effect of a faster mutation rate. For example, researchers have calculated that "mitochondrial Eve"—the woman whose mtDNA was ancestral to that in all living people—lived 100,000 to 200,000 years ago in Africa. Using the new clock, she would be a mere 6000 years old." The article also noted that the new findings of faster mutation rates pointing to mitochondrial Eve about 6,000 years ago have even contributed to the development of new mtDNA research guidelines used in the forensic investigations "adopted by the FBI."[245] Now, over 17 years later, and using even more human mtDNA data, Dr. Jeanson is spectacularly confirming this previously unheralded discovery.

The combined results of all these different genetic studies fit perfectly with the predictions of a biblical creation, complete with its recent timeframe for creation as provided in the Bible. The unbiased genetic clocks simply cannot have been ticking for millions of years.

In addition, evolution predicts a net gain of information over time, accompanied by natural selection removing harmful genetic variants. But instead, we see a human genome filling up with harmful genetic variants in every generation. Information loss or genetic entropy rules over all genomes. Clearly, the predictions based on Scripture align well with the discoveries being made in the field of genetics.

In the case of humankind, the Bible indicates that Adam and Eve were originally created with pristine error-free genomes. No harmful mutations were present. Then sin entered into the world at the point of man's rebellion against God in the Garden of Eden and the whole of creation became cursed and subject to futility as a result of man's sin. The human genome has essentially been on a steep downhill slide ever since this key point in time. We are not gradually evolving better and improved genomes through random processes. Instead, the recently measured genetic patterns of degradation clearly match the biblical model and timeframe given to us in the Scriptures.

There is no valid science behind human macro-evolution, but well documented empirical science supports biblical creation. This means that you and I have Adam, not apes, in our past. It also means we can trust the Bible's history and other teachings.

CHAPTER 7

The Problem of Over-design for Darwinism[246]

Jerry Bergman, Ph.D.

O ver-design refers to parts and potentials of our bodies that exist well beyond those required for mere survival. Typical abilities and skills achieved by the human brain, such as in music, math, memory, and design, are examples that illustrate over-design. In this chapter, we'll review these examples and other exceptional abilities that show extreme cases of mental over-design, providing strong support for intelligent design that cannot be explained by Darwinism. We will also show that the scientific literature that discusses human abilities also concludes that many human skills remain unexplained by evolution theory. This even includes Alfred Russel Wallace (the co-founder of the modern neo-Darwinism theory of evolution), who is convinced that over-design requires an intelligent creator.

In this chapter, over-design is used in a positive sense. Over-design can be a positive or negative in the field of design. In a negative sense, it means over-engineered for a given purpose that interferes with function. In a positive sense, it means design beyond that which is required and which is functional.[247]

One example of over-design is that the general strength (or capacity) of most human body organs and structures is considered by many anatomists to be far beyond that which is normally required for survival.[248] The fact that humans have pairs of certain organs (such as lungs and kidneys) and numerous organ back-up systems, such as the human spleen, that are normally not required for life, is clear evidence for over-design. University of California Medical School Professor Jared Diamond noted that research has determined our intestinal capacity is double that required for a healthy life, our kidney system is three times as large as required, and our pancreas is fully ten times the necessary size.[249]

Evolution would not have evolved structures by the "survival of the fittest" mechanism that remain largely unused for most of the entire population. Although mindless, the process of natural selection "selects" only for those structures that aid in an organism's survival, primarily as reflected in the animal's reproductive success. In Fisher's[250] words:

> Are you the richest man in America, the most powerful woman in business, the smartest kid in the class? Nature doesn't care. When Darwin used the term "survival of the fittest" he wasn't referring to your achievements or your endowments. He was counting your children. You may have flat feet, rotten teeth, and terrible eyesight, but if you have living children you are what nature calls "fit." You have passed your genes to the next generation, and in terms of survival you have won.

For this reason, Darwin asserted that traits not directly or indirectly contributing to a greater number of offspring would not be selected. In the words of genetics Professor Steve Jones, "Natural selection is no more than a machine. What it makes depends on what it has to work with and where it started."[251] The process of natural selection has no agency—no purposeful intelligence that can create or design with intention. The fact that many human organs and structures are far larger or more developed than required for survival is a major problem for Darwinism.[252] Two kidneys may enable some persons to live longer due to gradual kidney loss from aging, but this life extension advantage typically occurs long past reproduction age and, consequently, would not be selected.

As Ornstein concluded, "A major mystery in human evolution concerns why there is such a gigantic jump between the brains of *Homo habilis* and *Homo erectus* [modern humans]," adding that Alfred Russel Wallace, the co-discoverer of Darwin's theory, concluded that "the human brain was over-designed ... thus could not have evolved."[253] The case for over-design was so significant for Wallace that he concluded the human brain could not have evolved, but rather that it must have been created. The problem of over-design of the brain led Wallace to abandon the Darwinian theory of evolution, and to embrace what is today known as Intelligent Design.[254]

Wallace wrote, "The brain of prehistoric and of savage man seems to me to prove the existence of some power, distinct from that which has guided the development of the lower animals through their ever-varying forms of being."[255] His conclusion was that the brain's over-design demonstrated to him that "a superior intelligence has guided the development of man in a definite direction, and for a special purpose."[256] Thus, "he scientifically departed from Darwin over the evolution of the human brain, which Wallace could not conceive as being the product of natural selection... and thus must have been designed by a higher power."[257] This view of Wallace's was shared by many other naturalists of his day, and was the primary factor that caused a major rift between him and Darwin.[258]

Others argue that the over-designed extra capacity aids in cases of sickness or parasite infestation. This may be true in certain situations, but in most cases, overall health is far more critical in facilitating survival than kidney size. The evolutionary cost of maintaining excess kidney capacity, intestine length and other over-design luxuries would not compensate for the gain resulting from their potential advantages in rare situations. Excess capacity would normally be selected against, but the fact is that over-design—often called "redundancy" in biology—is everywhere. Researchers who remove certain genes in laboratory animals have found that the subjects either die young or, more commonly, because of over-design, they live blithely on even without what seemed to be essential parts of their machinery, including a gene for collagen, the structural material of much of the body, or certain genes that passes signals around the cell. Duplication is a useful insurance policy against the wiles of geneticists; but how and why these extra copies evolved, nobody knows.[259]

OVER-DESIGN IN THE HUMAN BRAIN

Harvard Professor Steven Pinker notes: "The human mental processes work so well we tend to be oblivious to their fantastic complexity, to the awe-inspiring design of our own mundane faculties. It is only when we look at them from the vantage point of science and try to explain their workings that we truly appreciate the nobility, the admirability, the infinite capacity of human faculties."[260] Because "we tend to be oblivious to their fantastic complexity," a study of those with exceptional mental abilities best illustrates the concept of over-design that convinced Wallace to accept intelligent design.

A prime example of over-design, the human brain, is an organ able to multitask, as do computers, but the brain has capabilities far exceeding those of man-made computers.[261] For example, the human brain is able to store between 100 trillion and 280 quintillion bits of information in three pounds of matter. It is protected by bony armor, cushioned by fluid, and serviced by a complex network of blood vessels. Everything about it exemplifies purpose and design. Life cannot begin without a brain, and life ends in four minutes without it.[262]

Although most of the brain consists of structural neurons and glial cells, these accessory structures are required for it to function. A brain design that allows learning new material for a length of time beyond even that of the most extreme expected human lifespan would not be "selected" by evolution. Rather, evolution theory holds that only the traits required for survival through the childbearing years would be selected. It is well recognized that our brain gave us a huge survival advantage for inventing, hunting, building, farming, cooking, treating illness, and even governing. The problem for evolution is that it was not until the twentieth century that any use existed for the phenomenal capacity of the human brain to perform such higher mathematical functions as nonlinear tensor calculus, relativistic quantum theory, and higher dimensional geometry. These abilities come at a cost: thirty-five percent of the entire blood flow in the human body services the brain. Moreover, to make room for the brain lobes that support mathematics, logic, analysis, communication, and meditation, the lobes that support some of our senses (smell and sound in particular) and of our muscles were reduced.[263]

Darwinism has always had difficulty accounting for the many documented examples of over-design, such as the fact that our brain is able to do higher mathematics, analysis, and meditation beyond the demands of mere survival. These three anticipatory endowments, among others, equip humanity for peak performance in a high technology environment. Unlike any other species of life, humans appear to have been equipped in advance for a life far different from the one they experienced when they first appeared. Such equipping of humanity, while puzzling from a Darwinian view, points to a Creator with foresight and a special plan for the creatures who bear His image.[264]

Simmons[265] notes that these mental traits include skills such as becoming a "gifted" artist. If this skill was "merely the luck of the genetic lottery," one must wonder why so many people have artistic talent, far more than employment as artists can support. Thus the source of the phrase "starving artist." He reasoned that the ability to paint a water scene, sing an aria, win a debate, or test a complex hypothesis cannot easily be attributed to survival of the fittest. The human race has gone beyond competition for mere survival. If one were to assume that gifts came about by a genetic accident, then one would have to explain how millions, if not billions, of extremely compatible neurological changes simultaneously came about in the brain and spinal cord.

Simmons notes that a major problem with Darwin's theory is its inability to explain the many phenomenal mental gifts that humans possess, such as laughing, singing, dancing, reading, playing, understanding, complex thinking, offering sympathy, and simply smiling. Experts on evolution rarely tackle these qualities because they can't explain them (by Darwinism). Is Albert Einstein a product of natural selection, or is he merely a product of many genetic mutations? Or Mark Twain? Or Gandhi? Or Shakespeare? Or Mother Teresa? What about "idiot savants" who can play thousands of songs on the piano without a lesson? Evolutionary theories do not explain these special skills.[266]

Mental over-design includes those who have exceptional abilities in math, music and other exclusively human endeavors.[267] Wallace concluded that: "Natural selection can only fashion a feature for immediate use. The brain is vastly overdesigned for what it accomplished in primitive society; thus, natural selection could not have built it."[268] Indeed, the sophistication and "extra" capabilities of our brains defy natural selection. Natural selection does not have the intelligence or intention to "fashion"— it is a mindless process that results in features that are both advantageous and disadvantageous for survival. We have so much more.

Part of Wallace's reasoning, which required a whole chapter in his 1870 book, was that a brain one-half larger than that of the gorilla would … have sufficed for the limited mental development of the savage; and we must therefore admit that the large brain he actually possesses could never have been solely developed by evolution, whose essence is, that they lead to a degree of organization exactly proportionate to the wants of each species, never beyond those wants…Natural selection could only have endowed savage man with a brain a little superior to that of an ape, whereas he actually possesses one very little inferior to that of a philosopher.[269]

Research scientist Dr. Frances Collins has been for several years the Director of the National Institutes of Health (NIH), which is the primary federal agency charged with conducting and supporting medical research. Dr. Collins concluded that over-design in humans, such as human language ability, self-awareness, and the ability to imagine the future, are all very convincing evidence of the creation of humans by God.[270] These were one of the central reasons that converted him from an atheist to a Christian.

Many specific examples of profoundly mentally gifted persons document over-design. Truman Henry Safford was a lightning calculator child prodigy born in 1936 in Royalton, Vermont. At the young age of ten, when asked by an examiner to square 365,365,365,365,365,365 in his head, Safford calculated the answer in seconds. Safford published his first almanac when he was nine, and many editions of his award-winning almanac were sold out. Safford graduated with honors from Harvard at the age of 18 after only two years of study.[271] Because he could remember the positions of all the stars listed in the Nautical Almanac, Safford was able to discover several new nebulae.[272] Although, like Isaac Newton he was described as a "sickly nervous child," he lived to age 65.[273]

Another example is the phenomenal memorizing and calculating abilities of savants that do not relate to survival.[274] (Savant is from Latin *sapere* meaning to be wise.) Calendar calculators can determine the day of the week for any date in history as well as any future date. Typical is one 20-year-old calculator, who, when asked questions such as what day of the week June 14, 1808 fell on, consistently correctly answered—in this case Tuesday, accurately compensating for the 1808 leap year.

Another skill is "instantaneous counting" where, for example, hundreds of matchsticks are dropped on a table and the counter can relate the correct number as soon as they hit the table.[275] Safford, the case noted above, could sweep his visual view across a long fence, and in seconds count every fence post as far as can be seen with the unaided eye—sometimes totaling in the hundreds. Yet another example is the ability to rapidly identify prime numbers, square roots, and other similar number feats.[276]

Further examples of over-design include trivia memorizers. Typical is the research of Columbia University's Harold Ellis Jones who extensively studied one trivial memorizer. To illustrate the abilities of his subject, Jones reported that the subject had successfully put to memory the following information:

1. The population of every town and city in the United States larger than 5,000 persons.
2. The county seats of all counties in the United States and the populations of about 1,800 foreign cities.
3. The distances of all cities in this country from New York and from Chicago, and also the distance from each city or town to the largest city in its state.
4. Statistics concerning close to 3,000 mountains and rivers.
5. The dates and essential facts connected with over 2,000 important inventions and discoveries.[277]

Examples of these amazing human abilities could continue for thousands of pages. In fact, it's likely that each person reading this book could identify several individuals who have superior abilities—abilities that are well beyond what's needed for just mere survival. But our human talents don't start and end with only intellectual abilities. They include "extra" athletic abilities too—a topic we'll explore next.

ATHLETIC TALENT

Many other examples of extraordinary gifts include athletes who exhibit a vast array of very accurate, rapid moves when playing their sport. This requires a high level of brain development, especially in the cerebellum part of the brain. Evolutionary ideas about "survival of the fittest" may be able to explain the ability to escape enemies or the skills needed to obtain food and shelter, but all of the skills discussed above are well beyond the level required for survival. As professor Niall Shanks admitted, evolution could have not equipped us to achieve tasks such as visualizing "four-dimensional objects in four-dimensional space-time" such as is required to fly an airplane at the speed of sound.[278] Yet, humans can do this and much more.

Ross explained the problem that over-design presents to the theory of evolution as follows:

> Human beings seem vastly "over-endowed" for hunter-gatherer or agrarian existence. For tens of thousands of years humanity carried intellectual capacities that offered no discernible advantage. From a Darwinian perspective, such capacities would be unlikely to arise and, even if they had randomly emerged, they would likely have been eliminated or minimized by natural selection.[279]

He notes that these capacities can easily be explained from a design perspective. Only today in our complex technological society do they serve the highly specialized needs of a technological society, benefiting the life quality and longevity of all humanity. The dexterity of the human hand certainly gave the human race an early survival advantage. Humans could craft more elegant tools and weapons than other bipedal primate species. However, the ability to type faster than a hundred words per minute seems to have offered no particular survival advantage until the twentieth century. Likewise, the remarkable ability to play a Liszt piano concerto had no utility until the invention of the piano.[280]

DARWINISTS ATTEMPT TO EXPLAIN OVER-DESIGN

Many Darwinists attempt to explain mental over-design by arguing that the evolution of human consciousness had a major downside: a conscious mind is aware of death, sickness and the cruelties existing in life. Music and art, therefore, exist to help one cope with these universal events. If this claim is true, it would seem that human consciousness would be adversely selected long before the complex skills needed to produce art and music developed to overcome the adverse results of consciousness. It would also appear that a far simpler way to cope with the survival disadvantage that a conscious mind produces would be selection for the ability to effectively accept the human condition.[281] Actually, as World War II and numerous holocausts have documented, many (but not all) humans eventually adjust somewhat to the suffering that results from active military combat, or dictatorial regimes.

Others argue that abilities, such as musical talent, are not directly related to survival, but rather they evolved because they facilitated social bonding and, therefore, indirectly aided in survival. Parsimony postulates that selection would favor social bonding itself, and not for a highly indirect means of achieving this goal, such as selecting for the ability to produce certain sound sets that we call music and also selecting for the mental ability to value more easily made sounds over those requiring significant talent.

If selection explained music and art talent, it would also select for a greater ability to enjoy the more common musical ability levels instead of the rare skills needed to produce a Mozart or Bach music level quality. The fact that many animals can also be taught human skills (such as chimps can learn mechanical typing skills) also does not argue against the over-design thesis, but rather it supports the view that animals also exhibit evidence of over-design.

Many Darwinists assume that over-design must be a result of mutations selected by natural selection, but many researchers recognize that this does not explain cases, such as savant syndrome. Biology Professor Stanley Rice concluded "High levels of intelligence evolved in the few species that do have it as a result of special circumstances that are still not understood."[282] The ability to mentally square 34,178,258 in seconds is not a skill that would facilitate survival in Africa where humans are believed to have evolved.[283] These skills are also so far above the norm that, if they conferred a clear survival advantage on them, they would be far more common.

Some evolutionists admit that, when attempting to explain the physiological cause of the savant syndrome, "most evolutionary reasoning remains at that qualitative, gee-whiz level and hasn't progressed since Darwin's day."[284] In other words, over-design is currently unexplainable from a Darwinist worldview and, as noted above, actually contradicts Darwinism. In an extensive review of the published literature, we have yet to locate even a remotely plausible explanation for mental over-design within the framework of natural selection. Most academic studies avoid the topic.[285]

CONCLUSION

Some argue that the idea of humans not using their full mental abilities is a myth. The fact is, few of us achieve the mental feats similar to the examples reviewed above. The finding that certain persons can achieve the mental and learning level of those discussed above, indicates that many more persons could also achieve at this level. These cases also illustrate that humans have abilities to engage in activities that go far beyond and outside the requirements needed for survival.[286] As Dembski concluded, "evolutionary process unguided by intelligence cannot adequately account for the remarkable intellectual and moral qualities that we see exhibited among humans."[287] Hawking added that: "scientists still cannot satisfactorily explain why ... so much human activity operates at a subliminal level. The spiritual sophistication of musical, artistic, politic, and scientific creativity far exceeds that of any primitive function programmed into the brain as a basic survival mechanism."[288]

The incredible many abilities of the brain argues that "the brain seems overdesigned, a feature that could not come about by evolution."[289] This observation is elegantly illustrated by the examples cited in this paper. Dembski also postulates that cases, such as William Sidis, are strong evidence that, in contrast to Darwin's conclusions, "the difference between humans and other animals is radical and represents a difference in kind and not merely a difference in degree" as Darwin claimed.[290] The fact of brain over-design "raises a fundamental problem for evolutionary theory. There is no reason why an evolved mind should have a capacity far beyond what most of us utilize" in our daily life.[291]

CHAPTER 8

Vestigial Structures in Humans and Animals

Jerry Bergman, Ph.D.

Most people have heard the common assertion that human bodies have some parts that are "leftover" from the evolutionary process that took "millions of years." Body parts such as the tailbone, tonsils, and the appendix are commonly placed in this category of "extra" or "unnecessary" body parts.

While many evolutionists are just fine with this assumption, many Christian's might ask, "Why would God—who is able to design humans in a complete and perfect fashion—leave such 'extra' or 'unnecessary' parts?" This question is answered by this section by explaining that these supposedly "extra" parts are not extra at all. We do this by providing current medical research that demonstrates just how intentional God was when He designed the human body.

INTRODUCTION

One major supposed proof of evolution is the observation that some organs appear to be degenerate or useless, often called vestigial organs. As Professor Senter opines, the "existence of vestigial structures is one of the main lines of evidence for macroevolution."[292] Vestigial organs are usually defined as body structures that were believed to have served some function in an organism's evolutionary history, but they are now no longer functional, or close to functionless.[293]

Thus, evolutionists teach that "living creatures, including man, are *virtual museums of structures that have no useful function*, but which represent the remains of organs that once had some use"[294] (emphasis added). Because all of the claimed vestigial organs have now actually been shown to be useful and integral to human function, evolutionists who attempt to salvage their idea have tried to shift gears. They now suggest that some organs have "reduced function," compared to their function in some undefined past. Thus, a new definition for "vestigial" is being used by some evolutionists. A problem with the revisionist definition is: Just how much reduction is required before the "vestigial" label is appropriate? Is 30% a large enough reduction, or will a 10% reduction suffice? In addition, there are so many putative examples of "reduced size" functional structures that the label "vestigial" becomes meaningless.

For example, an analysis of skull shapes of our supposed evolutionary ancestors shows that our human jaw is vestigial compared to our alleged ancestors, since it is claimed to be much smaller in humans today (and also has a reduced function relative to its strength and ability to chew food).[295] Furthermore, not only the human jaw and nose, but our eyes, eyebrows, front limbs, ears, and even our mouth could also be labeled vestigial when compared to our alleged ancestors. For this reason, the term becomes meaningless when defined in this fashion. Anything could be "vestigial" if it simply suits the writer.

Darwin discussed this topic extensively, concluding that vestigial organs speak "infallibly" to evolution.[296] Darwin asserted that the existence of vestigial organs is strong evidence against creation, arguing that vestigial organs are so "extremely common" and "far from presenting a strange difficulty, as they assuredly do on the old doctrine of creation, might even have been anticipated in accordance with evolution."[297]

The view that vestigial organs are critical evidence for macroevolution was further developed by the German anatomist Wiedersheim, who made it his life's work.[298] Wiedersheim compiled a list of over one hundred vestigial and so-called "retrogressive structures" that occur in humans. His list included the integument (skin), skeleton, muscles, nervous system, sense organs, digestive, respiratory, circulatory and urogenital systems.[299] Most of these remnants of (past physical) structures are found completely developed in other vertebrate groups.[300] Therefore, Wiedersheim concluded that the "doctrine of special creation or ... any teleological hypothesis" fails to explain these organs.[301]

For the medically-informed reader, we left most of the technical language in this chapter intact. Readers without this background, however, should still be able to read this chapter and gain an understanding that

God has an incredible design for each and every part of the human body!

VESTIGIAL PROBLEMS IN THE TEXTBOOKS

Let us now examine the most common vestigial organ claims. We hope your appreciation grows for God Who did in fact know what He was doing when He *created us in His image* (Genesis 1:27) and Who ensured we are *fearfully and wonderfully made* (Psalm 139:14).

THE COCCYX (TAILBONE)

Humans lack a tail. All lower primates have tails and the human coccyx (tailbone) is interpreted by Darwinists as a rudimentary tail left over from our distant monkey-like ancestors that supposedly had tails. Specifically, Darwin claimed that the "coccyx in man, though functionless as a tail, plainly represents this part in other vertebrate animals."[302]

A major problem with the conclusion that the coccyx shows evolution is that our supposed "nearest relatives" including chimpanzees, gorillas, orangutans, bonobos, gibbons or the lesser apes such as siamangs all lack tails! Only a few of the over one hundred types of monkeys and apes, such as spider monkeys, have tails. The primates that have tails tend to be the small cat-like lemurs and tarsiers.

In fact, the coccyx "is merely the terminal portion of the backbone. After all, it does have to have an end!"[303] The major function of the coccyx is an attachment site for the interconnected muscle fibers and tissues that support the bladder neck, urethra, uterus, rectum, and a set of structures that form a bowl-shaped muscular floor, collectively called the pelvic diaphragm, that supports digestive and other internal organs.[304]

The muscles and ligaments that join to the coccyx include the coccygeus muscle ventrally and the gluteus maximus muscle dorsally. The coccygeus muscles enclose the back part of the pelvis outlet.[305] The levator ani muscles constrict the lower end of both the rectum and vagina, drawing the rectum both forward and upward.[306] The cocygeus muscle, which is inserted into the margin of the coccyx and into the side of the last section of the sacrum, helps to support the posterior organs of the pelvic floor. The coccygeus muscle is a strong, yet flexible, muscle, often described as a "hammock," that adds support to the pelvic diaphragm against abdominal pressure. The coccyx muscle system expands and contracts during urination and bowel movements, and it also distends to help enlarge the birth canal during childbirth.[307]

Another useful structure connected to the coccyx is the anococcygeal raphe, a narrow fibrous band that extends from the coccyx to the margin of the anus.[308] Without the coccyx and its attached muscle system, humans would need a very different support system for their internal organs requiring numerous design changes in the human posterior.[309] Darwin was clearly wrong about the coccyx, and it is way past time that textbooks reflect known science about the well-designed end of the human spine.

THE TONSILS AND ADENOIDS

Among the organs long considered vestigial are the tonsils and adenoids. The tonsils are three sets of lymph tissues. The first, called palatine tonsils or "the tonsils," consist of two oval masses of lymph tissue (defined below) attached to the side wall at the back of the mouth. The second pair is the nasopharyngeal tonsils, commonly called the adenoids. The last section contains the lingual tonsils, which consist of two masses of lymph tissue located on the dorsum of the tongue.

The assumption that the tonsils are vestigial has been one reason for the high frequency of tonsillectomies in the past. Decades ago J. D. Ratcliff wrote that "physicians once thought tonsils were simply useless evolutionary leftovers and took them out thinking that it could do no harm. Today there is considerable evidence that there are more troubles of the respiratory tract after tonsil removal than before, and *doctors generally agree that simple enlargement of tonsils is hardly an indication for surgery*"[310] (emphasis added).

In recent years, researchers have demonstrated the important functions of both the tonsils and adenoids. As a result, most doctors are now reluctant to remove either the tonsils or the adenoids. Medical authorities now actively discourage tonsillectomies.[311]

The tonsils are lymph glands. They help establish the body's defense mechanism that produces disease-fighting antibodies. These defense mechanisms develop during childhood, as children sample and record materials through their mouths. The tonsils begin to shrink in the preteen years to almost nothing in adults, and other organs take over this defense function.[312] Because tonsils are larger in children than in adults, the tonsils are important in the development of the entire immune system.[313] For example, one doctor concluded that:

> The location of the tonsils and adenoids allows them to act as a trap and first line of defense against inhaled or ingested bacteria and viruses. The tonsils and adenoids are made up of lymphoid tissue which manufactures antibodies against invading diseases. Therefore, unless there is an important and specific reason to have the operation, it is better to leave the tonsils and adenoids in place. [314]

The tonsils are continually exposed to the bacteria in air we breathe and for this reason can readily become infected. As part of the body's lymphatic system, they function to fight disease organisms.[315] The tonsils "form a ring of lymphoid tissue" that guards the "entrance of the alimentary [digestive] and respiratory tracts from bacterial invasion." Called "super lymph nodes" they provide first-line defense against bacteria and viruses that cause both sore throats and colds.[316] Although removal of tonsils obviously eliminates tonsillitis (inflammation of the tonsils), it may increase the incidence of strep throat, Hodgkin's disease, and possibly polio.[317] Empirical research on the value of tonsillectomies in preventing infection demonstrate that the "tonsillectomy is of little benefit after the age of eight when the child's natural defenses have already made him immune to many infections." [318]

Just like calling the coccyx a useless evolutionary leftover, calling tonsils useless vestiges of organs that were only useful in our supposed distant evolutionary ancestor's bodies totally ignores the facts. These organs are well-designed and useful, just as if God created them on purpose.

THE VERMIFORM APPENDIX

The appendix was one of the "strongest evidences" used by Darwin to disprove creationism in his *The Descent of Man* (1871) book: "in consequence of changed diet or habits, the caecum had become much shortened in various animals, the vermiform appendage [appendix] being left as a rudiment of the shortened part... Not only is it useless, but it is sometimes the cause of death ... due to small hard bodies, such as seeds, entering the passage and causing inflammation." [319] Since Darwin, this claim has been repeated often in books and journals. The appendix was once commonly cited in many biology texts as the best example of a vestigial organ. [320]

The human appendix is a small, narrow, worm-shaped tube that varies in length from 1 to 10 inches.[321] Its average length is slightly over three inches long, and it is less than 1/2 inch wide.[322] The small intestine empties into the large intestine above the floor of the cecum at an entrance passage controlled by a valve. The lower right end of the large intestine in humans terminates somewhat abruptly at an area termed the cecum. The vermiform appendix is connected to the lower part of the cecum.

THE SAFE HOUSE ROLE

Most bacteria in a healthy human are beneficial and serve several functions, such as to help digest food. If the intestinal bacteria are purged, one function of the appendix is to replenish the digestive system with beneficial bacteria. Its location—just below the normal one-way flow of food and germs in the large intestine in a sort of gut cul-de-sac—supports the safe house role by protecting and fostering the growth of "good germs" needed for various uses in the intestines and enabling the digestive bacteria system to "reboot" after bouts of disease such as cholera or the use of antibiotics. Diarrhea is designed to flush out all bacteria from the colon, both good and bad. The bacteria in the appendix are not affected by diarrhea and can rapidly repopulate the colon to quickly reestablish healthy digestion.

For years, we noticed few effects of removing the appendix. Evolutionists thought that if people don't need them, they must be useless. And if it's useless, then it must be a remnant of some evolutionary ancestor that did need it for something. However, just because removing a body part does not immediately kill you does not mean that it has no use. One can lose the end of some fingers and still do almost everything that fully-fingered people do, but fingertips are still useful. Like fingertips, tonsils and the appendix are useful, and as far as is known, they always have been ever since God created them.

THE FUNCTIONS OF THE APPENDIX IN DEVELOPMENT

The appendix is also involved in producing molecules that aid in directing the movement of lymphocytes to other body locations. During the early years of development, the appendix functions as a lymph organ, assisting with the maturation of B lymphocytes and in the production of immunoglobulin A (IgA) antibodies. Lymph tissue begins to accumulate in the appendix soon after birth and reaches a peak between the second and third decades of life. It decreases rapidly thereafter, practically disappearing after the age of about 60.

The appendix functions to expose white blood cells to the wide variety of antigens normally present in the gastrointestinal tract. Thus, like the thymus, the appendix helps suppress potentially destructive blood- and lymph-borne antibody responses while also promoting local immunity.[323]

In summary, researchers have concluded, "Long thought to be an evolutionary remnant of little significance to normal physiology, the appendix has ... been identified as an important component of mammalian mucosal immune function, particularly B lymphocyte-mediated immune responses and extrathymically derived T lymphocytes."[324] Calling the appendix "vestigial" is a big mistake.

THE THYROID

The thyroid is a two-lobed gland connected by a narrow strip located just below the voice box.[325] German Darwinist Ernst Haeckel long ago asserted that not only is the thyroid vestigial, but that our body contains "many rudimentary organs.... I will only cite the remarkable thyroid gland (thyreoidea)."[326] Because surgeons found that adults could survive after having their thyroid removed, it was assumed by some that it was useless. Wiedersheim listed the thyroid as vestigial because of the "manner in which the thyroid originates."[327] Were they right? Modern medicine has revealed enough about the thyroid for us to find out.

The thyroid is one of the largest endocrine glands, and can grow to as large as 20 grams in adults. The three most important hormones it produces are triiodothyronine (T3) and thyroxine (T4), both of which regulate metabolism, and calcitonin, which regulates calcium levels. Both T3 and T4 stimulate the mitochondria to provide more energy for the body and increase protein synthesis. Without T3 and T4, humans become sluggish, and growth stops. An oversupply (or an undersupply) of thyroxine results in over-activity (or under-activity) of many organs. Defects in this organ at birth can cause a deformity known as cretinism, shown as severe retardation of both physical and mental development.[328] Haeckel was exactly wrong about the Thyroid, but he didn't know its values. Museums and textbook displays still portraying the thyroid as vestigial show an almost criminal disregard of good observational science.

THE THYMUS

The thymus gland is an example of an important organ that was long judged not only vestigial, but harmful if it became enlarged. Maisel reported that for generations physicians regarded it "as a useless, vestigial organ."[329] Clayton noted that an oversized thymus was once routinely treated with radiation in order to shrink it.[330] Follow-up studies showed that, instead of helping the patient, such radiation treatment caused abnormal growth and a higher level of infectious diseases that persisted longer than expected.

The thymus is a small pinkish-gray body located below the larynx and behind the sternum in the chest.[331] A capsule, from which fingers extend inward, surrounds it and divides it into several small lobes, each of which contains functional units called follicles.

FUNCTIONS OF THE THYMUS

This once-deemed worthless vestigial structure is now known to be the master gland of the lymphatic system. Without it, the T-cells that protect the body from infection could not function properly because they develop within the thymus gland. Researchers have now solved the thymus enigma, finding that far from being useless, the thymus regulates the intricate immune system which protects us against infectious diseases. Thanks to these discoveries, many researchers are now pursuing new and highly promising lines of attack against a wide range of major diseases, from arthritis to cancer.[332]

The cortex, or outer tissue layer, of the thymus is densely packed with small lymphocytes surrounded by epithelial-reticular cells. The lymphocytes, also called thymic cells, are produced in the cortex and exit the gland through the medulla.[333] The medulla is more vascular than the cortex, and its epithelial-reticular cells outnumber the lymphocytes.

Besides being a master regulator and nursery for disease-fighting T-cells, the thymus takes a dominant role in reducing autoimmune problems. These occur where the immune system attacks the person's own cells, which is called the self-tolerance problem.[334] As research on immune tolerance continues, "the multiplicity of mechanisms protecting the individual from immune responses against self-antigens" and "the critical role the thymus plays is becoming better understood."[335] "Evidence now exists that regulatory cells have a role in preventing reactions against self-antigens, a function that is as important as their role of clonal deletion of high-affinity self-reactive T-cells."[336]

Regulatory T-cells also help to prevent inappropriate inflammatory responses to non-disease-causing foreign antigens. This system plays an essential role in preventing harmful inflammatory responses to foreign antigens that come in contact with mucous membranes, such as in many allergies.

In summary, a primary function of the thymus is to nurse to maturity small white blood cells called lymphocytes, which are then sent to the spleen and the lymph nodes, where they multiply.[337] There is nothing vestigial about the thymus.

THE PINEAL GLAND

The pineal gland was first described by French psychiatrist Philip Pineal in the 1790s.[338] The pineal body is a cone-shaped gland positioned deep inside the head, near the brain stem. Scientists are now finding out that the pineal gland's functions include regulating hormones:

> Scientists are closing in on a mystery gland of the human body, the last organ for which no function has been known. It is turning out to be a lively performer with a prominent role in the vital hormone producing endocrine system… Medical science is now finding what nature really intended by placing a pea-sized organ in the middle of the head.[339]

Of course, the Creator really deserves credit for the pineal gland, not nature. Nevertheless, the pineal gland also serves in reproduction:

> It has long been known that reduction in the amount of light reaching the eyes stimulates this small gland to synthesize and secrete an anti-gonadotrophic hormone(s) which results in marked attenuation of virtually all aspects of reproductive physiology.[340]

Researchers at the National Institute of Mental Health found that the pineal gland is a very active member of the body's network of endocrine glands, especially during certain growth stages.

THE PINEAL GLAND AND MELATONIN PRODUCTION

The pineal gland's most commonly mentioned function is its role in producing the hormone melatonin.[341] Cells in the pineal gland produce a special enzyme that converts serotonin to melatonin.[342] Melatonin is produced mainly in the pineal gland of vertebrates, but is also produced in a variety of other tissues. [343]

Light-dark levels are communicated to the brain from the retina to the pineal gland and help regulate melatonin levels. Melatonin is also a sleep-inducing hormone. This is why darkness generally promotes sleepiness.[344]

Melatonin also has important immune function stimulatory properties. It enhances the release of T-helper cell type 1 cytokines such as gamma-interferon and IL-2, counteracts stress-induced immunodepression and other secondary immunodeficiencies, protects against lethal viral encephalitis, bacterial diseases, and septic shock, and diminishes toxicity associated with several common chemotherapeutic agents.[345] The administration of melatonin also increases thymus cellularity and antibody responses.[346] Conversely, pinealectomy both accelerates thymic involution and depresses the humeral and cell-mediated immune response.[347]

PINEAL AND REPRODUCTION

The pineal gland is the primary controller of the timing of the onset of puberty, a critical developmental function. Melatonin regulates the production of anti-gonadotropin hormones. These help block the effects of hormones that stimulate gonad development. Damage to the pineal gland leads to early puberty in males. Conversely, if the pineal gland is overactive, puberty is delayed. Among melatonin's many other reproductive functions is regulation of the estrus cycle in women. Melatonin levels decrease as women age, particularly after they pass child-bearing age.[348] Changes in melatonin levels may be responsible for some sleep difficulties in menopausal females.

Before the advent of modern artificial lighting, the number of hours humans spent in darkness was much greater. Today, bright lighting found in almost all homes and offices may be affecting our reproductive cycle. Exposure to a large amount of light during most of one's waking hours may cause the onset of sexual maturity at an earlier age and even the higher rate of multiple births.

Studies on "pre-electric" Inuit Indians support the conclusion that light and the pineal gland are important in reproduction. When it is dark for months at a time in their arctic home, Inuit women stop producing eggs altogether and men become less sexually active. When daylight returns, both the women and the men resume their "normal" reproductive cycles.[349]

THE "NICTITATING MEMBRANE" IN THE HUMAN EYE

An excellent example of another commonly mislabeled vestigial organ is the so-called nictitating membrane remnant in the human eye. A nictitating membrane, or "third eyelid," is a very thin and transparent structure that small muscles move horizontally across the eye surface to clean and moisten the eye while maintaining sight. It hinges at the inner side of the lower eyelid of many animals. To nictitate means to move rapidly back and forth over the front of the eye.[350] The nictitating membrane is especially important in animals that live in certain environments, such as those that are exposed to dust and dirt like birds, reptiles, and mammals, or marine animals such as fish. Charles Darwin wrote about the "nictitating membrane:"

...with its accessory muscles and other structures, is especially well developed in birds, and is of much functional importance to them, as it can be rapidly drawn across the whole eye-ball. It is found in some reptiles and amphibians, and in certain fishes, as in sharks. ... But in man, the quadrumana, and most other mammals, it exists, as is admitted by all anatomists, as a mere rudiment, called the semilunar fold.[351]

Many continue to repeat Darwin's wrong idea about this membrane being a vestigial structure, even though, as we will show, it is clearly important in the human eye. [352]

ITS USE IN HUMANS

The classic eye anatomy textbook, *Clinical Anatomy of the Eye* by Snell and Lemp, accurately describes what we now recognize as the misnamed nictitating membrane. The plica semiluminaris, or "plica" for short, is a semilunar fold located on the inner corner of the eye which allows that side of the human eyeball to move further inward, toward the nose.[353] Its anatomy reveals a delicate half-moon-shaped vertical fold. The eye has about 50–55% rotation, but without the plica semilunaris, the rotation would be much less. There exists slack that must be taken up when the eye looks forward or side-to-side; hence the fold. No such arrangement exists for looking up or down, for at this area the fornix is very deep. The absence of a deep medial fornix is required for the puncta to dip into superficial strips of tear fluid.[354] Because the plica allows generous eye rotation, it actually is an example of over-design. [355]

Another function of the plica semilunaris is to collect foreign material that sticks to the eyeball. Stibbe notes that on a windy day the eyes can rapidly accumulate dust, but due to the plica they can usually effectively remove it.[356] To do this, it secretes a thick sticky fatty liquid that effectively collects foreign material and, in essence, insulates the material for easy removal from the eye without fear of scratching or damaging the delicate eye surface. The critical role of the plica in clearing foreign objects from the eye surface has been recognized since at least 1927. This should be an embarrassment to those who have thought of it as vestigial since then.

MUSCLE AND BONE VARIATIONS AS VESTIGIAL ORGANS

Most of the over one hundred vestigial organs and structures listed in Wiedersheim's original 1895 work were small muscles or minor variations in bones and were not glands or discrete organs, such as the human appendix.[357] Many of these muscles were labeled vestigial because they were small and made only a small contribution (or supposedly no contribution) to the total muscle force. The problem is that if a muscle was vestigial it would rapidly shrink, as research on living in a weightless situation, such as in outer space, has documented.

Thus, if a muscle has not atrophied it must be functional. It is now known that most small, short body muscles produce fine adjustments in the movement of larger muscles, or serve other roles, such as in proprioception.[358] The proprioceptive system allows the body to rapidly and accurately control limb position. It is why falling cats so often land on their feet. Anatomist David Menton concludes that:

> ...most muscles have a sensory function in addition to their more obvious motor function. ...that some of the smaller muscles in our body that were once considered vestigial, on the basis of their small size and weak contractile strength, are in fact sensory organs rather than motor organs.[359]

Certain other muscles and bone variations are also labeled vestigial primarily because they are not present in most (or many) people and are not required for survival. As is clearly evident in human skill differences, these muscle variations help to produce the enormous variety in many abilities so evident in modern humans. An example is the gross body muscle development of the stereotyped computer programmer compared with a football player. More commonly, many muscles are not well developed in most persons today in Western society due to our sedentary lifestyle.

This does not mean that they are vestigial, but only demonstrates their lack of use in modern life. It also demonstrates a very different lifestyle today than in the past. Lifestyle differences could cause many of these "less developed" muscles to be much larger. Would evolutionists have called them vestigial if they saw how much larger they were in a more athletic person's body? The fact that some individuals are superior athletes from a young age is evidence that genetic components clearly play an important role in complex physical activities. DeVries maintains that athletic ability depends on variations of numerous aspects of muscle cell structure and physiology.[360] Certain muscles and muscle types must first be present before they can ever be developed by proper training.

Gifted athletes, such as gymnastic and acrobatic stars, may tend to have certain muscles that some people may not even possess, or they can develop certain muscles to a greater extent. Most human abilities appear to be influenced by genetic differences that result from body structure variations. It follows that the human muscle system would likewise be influenced by heredity.

The argument that some small muscle is vestigial depends heavily on judgments as to the value and the individual use of a particular structure. It is clear that none of the so-called vestigial muscles are in any way harmful. Indeed, if they are developed at all, then those who have them may enjoy an advantage in certain activities, even if it is only an athletic or aesthetic advantage.

Scientists have clearly identified specific and well-designed purposes for every single supposedly vestigial organ so far proposed. Darwinist books, movies, and displays are dead wrong if they promote the concept of vestigial organs, which don't actually exist.

CHAPTER 9

Biblical Authority

Daniel A. Biddle, Ph.D.

C hristians who hold to *biblical authority* believe that the Bible was *written by God through man.* Thus, we believe the Bible to be true regarding all areas it speaks about—including history, the origin of the world, and the creation of the human race. We believe "God is not a man, that he should lie; neither the son of man, that he should repent" (Numbers 23:19—see also Titus 1:2 and Hebrews 6:18). Even though man is imperfect, we believe that God wrote the Bible through men as they were directed by God. "All Scripture is inspired of God and beneficial for teaching, for reproving, for setting things straight, for disciplining in righteousness, that the man of God may be fully competent, completely equipped for every good work" (2 Timothy 3:16–17).

Holding to biblical authority also means we believe that the line-up of patriarchs listed in Genesis Chapters 5 and 11, and repeated elsewhere in the Bible, leads back to Adam, the first created man at the end of the six-day creation week. The lifespans and birth years of the patriarchs can be used to calculate the time from Adam to Abraham (about 2,000 years) and from Abraham to Jesus (about 2,000 years), making about 4,000 years of history from Creation to Christ. Adding the time from now to Christ gives us the last 2,000 years, making about 6,000 years total. The chart below shows how clearly the first 11 chapters of Genesis show world history from Creation to about 2000 B.C. using the birth/death years listed from the patriarchs:

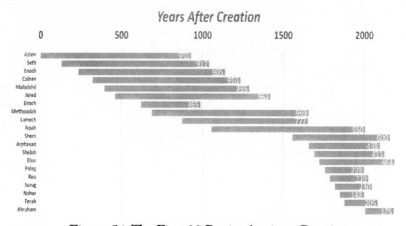

Figure 54. The First 20 Patriarchs since Creation

Next we'll review the current landscape of human origin beliefs in the U.S.

A BIBLICAL VIEW IN AMERICAN CULTURE

When it comes to the origins of mankind, the majority of Americans (regardless of religious orientation) hold to a "recent" view, with 46% believing that God created humans in their present form less than 10,000 years ago.

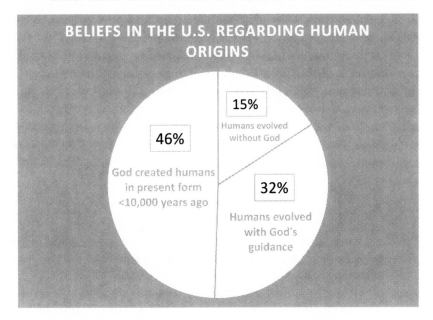

Figure 55. Beliefs in the U.S. Regarding Human Origins

Figure 55 comes from a survey conducted by a 2014 Gallup poll.[361] The pollsters conducted telephone interviews with a random sample of 1,028 adults, aged eighteen and older, living in all fifty U.S. states. The results of the survey show that more than 4 in 10 Americans continue to believe that God created humans in their present form less than ten thousand years ago, a view that has changed little during the past three decades.

Evolutionists scoff at the idea that the earth is this "young," but plenty of scientific and historical evidence supports the position. Ministries like *Answers in Genesis*, the *Institute for Creation Research*, and *Creation Ministries International* have amassed mounds of evidence that show how science backs the historical Genesis position. We intend to give excellent reasons why millions of Americans are exactly right about recent creation despite generations of dogmatic evolutionary teaching.

WAS ADAM A MYTH OR A REAL PERSON?

Secular and Christian evolutionists do not accept that Adam was a real, historical person. They assert he was just a "myth" included in the Bible to represent some idealized picture of humanity or perhaps the culmination in the evolution of people from ape-like ancestors. But this view squarely contradicts both the Old and New Testaments. For example, in the first book of the Bible, Adam is described as being the first created man who had a son Seth with Eve when he was 130 years old, became a grandfather when he was 235, and died when he was 930 (Genesis 5). This description does not sound like a myth, but factual history. The writers of the New Testament also believed that Adam was a real historical person, as shown by the Scriptures below, written about 3,000 years after Adam died:

- "Nevertheless, death reigned from the time of Adam to the time of Moses, even over those who did not sin by breaking a command, as did Adam, who is a pattern of the one to come" (Romans 5:14).
- "For just as through the disobedience of the one man the many were made sinners, so also through the obedience of the one man the many will be made righteous" (Romans 5:19).
- "So it is written, 'The first man Adam became a living being'; the last Adam, a life-giving spirit. The spiritual did not come first, but the natural, and after that the spiritual. The first man was of the dust of the earth; the second man is of heaven. As was the earthly man, so are those who are of the earth; and as is the heavenly man, so also are those who are of heaven. And just as we have borne the image of the earthly man, so shall we bear the image of the heavenly man" (1 Corinthians 15:45–49).
- Jesus referred to Adam and Eve as the first two humans who were created by God at the very beginning: "But from the beginning of the creation God made them male and female" (Mark 10:6).
- Adam's lineage is treated as literal history in the book of Jude: "Enoch, *the seventh from Adam*, prophesied about them: 'See, the Lord is coming with thousands upon thousands of his holy ones'" (Jude 14).
- The Gospel of Luke records Jesus' ancestry back to "The son of Enosh, the son of Seth, the son of Adam, the son of God" (Luke 3:38).
- "For Adam was formed first, then Eve. And Adam was not the one deceived; it was the woman who was deceived and became a sinner" (1 Timothy 2:13–14).

These verses clearly show that Adam was a real, historical person who lived, had a family, and died in a certain time in history (the *very beginning* according to Jesus). This truth is important to the Christian faith—in fact, it is fundamental, as the respected Welsh theologian Dr. Martyn Lloyd-Jones explained when he wrote,

The Bible does not merely make statements about salvation. It is a complete whole: it tells you about the origin of the world and of man; it tells you what has happened to him, how he fell and the need of salvation arose, and then it tells you how God provided this salvation and how He began to reveal it in parts and portions. Nothing is so amazing about the Bible as its wholeness, the perfect interrelationship of all the parts. Therefore, these early chapters of Genesis with their history play a vital part in the whole doctrine of salvation… Paul's whole case [in Romans 5] is based upon that one man Adam and his one sin, and the contrast with the other one man, the Lord Jesus Christ, and His one great act. You have exactly the same thing in 1 Corinthians 15; the apostle's whole argument rests upon the historicity. Indeed, it seems to me that one of the things we have got to assert, these days in particular—and it should always have been asserted— is that our gospel, our faith, is not a teaching; it is not a philosophy; it is primarily a history.[362]

Why do some people assert that Adam was a fiction? In short, they reason in a circle, assuming the conclusion from the beginning. They begin with the belief that humans evolved from ape-like ancestors. Adam has no role or place in the evolutionary story, so he must be fiction. Notice how they dismiss Adam without even examining the historical evidence. Now, what about Adam's wife, "the mother of all living" (Genesis 3:20)?

THE CREATION OF EVE, THE FIRST WOMAN

The Bible also clearly presents God's creation of the first woman (Eve), being drawn from the Adam's side:

So the Lord God caused a deep sleep to fall upon the man, and he slept; then He took one of his ribs [or "sides"] and closed up the flesh at that place. The Lord God fashioned into a woman the rib which He had taken from the man, and brought her to the man. The man said, 'This is now bone of my bones, and flesh of my flesh; she shall be called

Woman, because she was taken out of Man.' (Genesis 2:21–23)

God created Eve to *complete* Adam, and only together do they form "man" in the complete image of God. "He created them male and female, and He blessed them and *named them Man* in the day when they were created" (Genesis 5:2).

Some people believe that the book of Genesis is more figurative than literal history, suggesting that God used evolution to bring about modern humans. Dr. Martyn Lloyd-Jones explains that God's creation of Eve confronts this. As a former medical doctor before he became a small-town minister, evolution did not intimidate Lloyd-Jones. Instead of evolutionary stories, he saw that the Bible recorded the real history of mankind. He wrote,

> If you do not accept this [Genesis] history, and prefer to believe that man's body developed as the result of an evolutionary process, and that God then took one of these humanoid persons, or whatever you may call them, and did something to him and turned him into a man, you are still left with the question of how to explain Eve, for the Bible is very particular as to the origin of Eve. All who accept in any form the theory of evolution in the development of man completely fail to account for the being, origin, and existence of Eve.

Another challenge with believing that evolutionary processes generated Adam or Eve is the New Testament, which completes the Old Testament. It also teaches that God created Adam first, and then Eve. When the Apostle Paul wrote to the Corinthian church about the creation of man and woman, he said: "For man is not from woman, but woman from man. Nor was man created for the woman, but woman for the man" (1 Corinthians 11:8–9). He also said in 1 Timothy 2:13: "For Adam was formed first, then Eve." This agrees with the order in Genesis. Some tell stories of soulless humans called Pre-Adamites living for thousands of years before Adam, but this contradicts what the Bible that teaches. If we say that we believe the Bible to be the Word of God, we must say that about the whole of the Bible. When the Bible presents itself to us as history, we must accept it as history.

CONCLUSION

When showing a Christian friend some of the images from this book that demonstrate how ape hips are perfectly designed for living in trees and walking on all fours and human hips are perfect for walking and running, he remarked "It's just so obvious that apes and humans were created differently by God! Why would you need to point that out to anyone?" That's a good question. If anyone opens their eyes and just looks around at the varieties of apes living today and the variability within humans, it's quite obvious that they are animals and humans are humans— with each reproducing "after their kind" just as forecasted by the Bible. There are no creatures living today that are between apes and humans, neither were there in the past. All that exists that "proves" evolution today is a pickup truck full of miscellaneous old bones that scientists desiring to believe in evolution have assembled in a story that "proves" it. The tales of Java Man, Piltdown Man, and Nebraska Man show just how much these faith-abandoning "scientists" want to leave behind the obvious creative acts of God and replace it with a self-assuring story of evolution they believe will get them off the hook from having to answer to an Almighty God after they die. It's both disturbing and sad.

Further, how is it that 96% of state education systems require evolution to be taught as fact, while 70% of Americans are "Christian" and 46% believe that God recently (and miraculously) created humans? Does that make any practical sense? Can you think of any other topic where half the population believes something about the past (like our origins) but almost all of educational systems teach the opposite? Surely this is true only in one area: our origins, including the questions of whether God exists and whether He has revealed Himself to us through history.

So what *is* going on with this topic in America? What explanation can make sense of all of this? Fortunately, the Bible gives us some clues. First, the Bible is clear that the systems of the world are under the "sway" or "control" of the enemy (1 John 5:19: "We know that we are of God, and the whole world lies under the sway of the wicked one"). We also know that in the end times a "strong delusion" will come that will lead people astray from God: "And for this reason God will send them strong delusion, that they should believe the lie, that they all may be condemned who did not believe the truth but had pleasure in unrighteousness" (2 Thessalonians 2:11–12). 2 Peter 3:3–6 also confirms that in the last days people will abandon the idea of the spontaneous supernatural creative and catastrophic acts of God, and use this philosophy to deny His sudden return.

All of this points back to the spiritual battle for our souls. The great lie of evolution convinces people there is no God and this will result in many entering eternity without Him. That's why this battle must be fought by each Christian, but in *sensitive and tactful ways*. Just ask an atheist friend what they think of "Bible believing fundamentalists." You will find that Christians have not done a good job at lovingly presenting the Truth. Rather, the reputations that Christians have developed is one where we are known for what we are *against*, rather than what we are *for*.

Sadly, this is quite the turn off to many seekers. Realize that walking up to a typical atheist and trying to convince them that God spontaneously created all life on earth just thousands of years ago will likely be too much to stomach—especially when they've received a lifetime of "millions of years of evolution" from almost every school, museum, and state park they've ever visited. This is why relationship building and prayer is so important. Once someone comes to faith, the "scales" will fall off of their eyes and they will be spiritually open to the truth about Creation.

But don't take this tact too far. Some hold to a strict "Jesus before Genesis" philosophy of evangelism, but in many cases, answering questions about Genesis is necessary *before* people will consider Jesus. When trying to "convert" the philosophical Greeks (who were culturally very similar to many people today), Paul first started with the fact that we are created by a loving God (Acts 17:26–27: "And He has made from one blood every nation of men to dwell on all the face of the earth, and has determined their pre-appointed times and the boundaries of their dwellings, so that they should seek the Lord, in the hope that they might grope for Him and find Him, though He is not far from each one of us"). Many people will likely want to explore the evidence for Biblical Creation (over the evolutionary alternative) before going further down the road towards becoming a Christian, and that's exactly why we wrote this book.

HELPFUL RESOURCES

The following websites are recommended for further research:

- Answers in Genesis: *www.answersingenesis.com*
- Answers in Genesis (High School Biology): *www.evolutionexposed.com*
- Creation Ministries International: *www.cmi.org*
- Institute for Creation Research: *www.icr.org*
- Creation Today: *www.creationtoday.org*
- Creation Wiki: *www.creationwiki.org*
- Evolution: The Grand Experiment with by Dr. Carl Werner: *www.thegrandexperiment.com*

ENDNOTES

1   Bill Bryson, *A Short History of Nearly Everything* (London: Black Swan Publishing, 2004), 529.

2   Unless, of course, scientists arrive at this forbidden place through "bio-hacking."

3   The phrase, "Let Us make man in Our image..." is a reflection of the Triune nature of God (Father, Son, Spirit) (see also Genesis 3:22 and 11:7).

4   The first 99% similarity claim, which Cohen calls "The Myth of 1%," was made in 1975 by Allan Wilson and Mary-Claire King using reassociation kinetics of single-copy DNA (J. Cohen, "Relative Differences: The Myth of 1%, *Science* 316 (2007): 1836. Similar studies came up with an average divergence in single-copy DNA that measured about 1.5%, producing the widely spread quotes of 98.5% DNA sequence similarity (B.H. Hoyer, et al., "Examination of Hominid Evolution by DNA Sequence Homology, *Journal of Human Evolution* 1 (1972): 645–649.

5   Jeffrey P. Tomkins, "Documented Anomaly in Recent Versions of the BLASTN Algorithm and a Complete Reanalysis of Chimpanzee and Human Genome-Wide DNA Similarity Using Nucmer and LASTZ," (October 7, 2015), Answers in Genesis: *https://answersingenesis.org/genetics/dna-similarities/blastn-algorithm-anomaly/*

6   Rod Preston-Mafham and Ken Preston-Mafham. *Primates of the World* (New York: Facts on File, 1992).

7   Credit: *http://www.godrules.net/evolutioncruncher/2evlch18a.htm*

8   Dr. Claud A. Bramblett, Professor Emeritus in the Department of Anthropology Department at the University of Texas.

9 Thomas Schoenemann. *An MRI Study of the Relationship Between Human Neuroanatomy and Behavioral Ability (Appendix A)*. Diss. University of California, Berkeley, 1997. (For comparison purposes, only the primates that had average body weights that exceeded 10 pounds were included).

10 Jeremy Taylor, *Not a Chimp: The Hunt to Find the Genes that Make Us Human* (Oxford University Press Oxford, UK, 2009): 217.

11 Schoenemann, 1997.

12 Schoenemann, 1997, Chapter 2.

13 J. Zeng, et al. "Divergent whole-genome methylation maps of human and chimpanzee brains reveal epigenetic basis of human regulatory evolution." *American Journal of Human Genetics* 91, 3 (2012): 455-465.

14 Brian Thomas, "Stark Differences Between Human and Chimp Brains," *Institute for Creation Research: www.icr.org/article/stark-differences-between-human-chimp/* (September 1, 2015).

15 James K. Rilling, "Human and Non-Human Primate Brains: Are They Allometrically Scaled Versions of the Same Design?" *Evolutionary Anthropology* 15 (2006): 65-77.

16 J.H. Balsters, E. Cussans, J. Diedrichsen, K.A. Phillips, T.M. Preuss, J.K. Rilling, N. Ramnani, "Evolution of the Cerebellar Cortex: The Selective Expansion of Prefrontal-projecting Cerebellar Lobules." *Neuroimage* 43 (2010): 388-98.

17 Rilling, *Human and Non-Human Primate Brains: Are They Allometrically Scaled Versions of the Same Design*, 69.

18 Taylor, 2009, 222.

19 K. Semendefer, A. Lu, N. Schenker, & H. Damasio, "Humans and Great Apes Share a Large Frontal Cortex," *Nature Neuroscience* 5 (3) (2002): 272–6.

20   E.A. Nimchinsky, E. Gilissen, J.M. Allman, D.P. Perl, J.M. Erwin, & P.R. Hof, "A Neuronalmorphologic Type Unique to Humans and Great Apes," *Proceedings of the National Academy of Sciences USA*, April 27, 96 (9) (1999): 5268-73.

21   Taylor, 2009, 234.

22   Taylor, 2009, 235.

23   Taylor, 2009, 236.

24   Taylor, 2009, 237.

25   Taylor, 2009, 245.

26   Sana Inoue & Tetsuro Matsuzawa, "Working Memory of Numerals in Chimpanzees," *Current Biology*, 17 (2007), R1004-R1005.

27   A Siberberg & D. Kearns, Memory for the Order of Briefly Presented Numerals in Humans as a Function of Practice, *Animal Cognition*, 12 (2009): 405–407.

28   P. Cook & M. Wilson, "In Practice, Chimp Memory Study Flawed," *Science*, 328 (2010): 1228; N. Humphrey, "This chimp will kick your ass at memory games – but how the hell does he do it?" *Trends in Cognitive Science*, 16 (2012): 353–355.

29   Fred Spoor, Theodore Garland, Gail Krovitz, Timothy M. Ryan, Mary T. Silcox, Alan Walker, "The Primate Semicircular Canal System and Locomotion," *Proceedings of the National Academy of Sciences* 104, 26 (2007): 10808–12.

30   P. Gunz, et al., "The Mammalian Bony Labyrinth Reconsidered: Introducing a Comprehensive Geometric Morphometric Approach," *Journal of Anatomy* 220, 6 (2012): 529-543.

31   Adam Summers, "Born to Run: Humans will Never Win a Sprint against your Average Quadruped. But our Species is well-adapted for the Marathon," *Biomechanics: www.naturalhistorymag.com/biomechanics/112078/born-to-run* (September 1, 2015).

32   Michael Tomasello, Brian Hare, Hagen Lehmann and Josep Call, "Reliance on Head Versus Eyes in the Gaze following of Great Apes and Human Infants: The Cooperative Eye Hypothesis," *Journal of Human Evolution* 52 (2007): 314-320.

33   Michael Tomasello, "For Human Eyes Only" (January 13, 2007), *New York Times: www.nytimes.com/2007/01/13/opinion/13tomasello.html* (September 1, 2015).

34   P. Ekman, & W. V. Friesen, *Facial Action Coding System*, (Palo Alto, CA: Consulting Psychologists Press, 1978).

35   A.M Burrows, B. M., Waller, Parr, L. A., & Bonar, C. J. (2006). Muscles of facial expression in the chimpanzee: Descriptive, ecological and phylogenetic contexts. *American Journal of Physical Anthropology*, 208, 153–167.

36   N.M. Schenker, W.D. Hopkins, M.A. Spoctor, et al, "Broca's area homologue in chimpanzees (Pan troglodytes): Probabilistic Mapping, Asymmetry, and Comparison to Humans," *Cereb Cortex* 20 (2010): 730-742.

37   Credit: *http://www.harunyahya.com/en/Books/4066/atlas-of-creation--/chapter/15490*

38   Smithsonian National Museum of Natural History, "What Does it Mean to be Human? Bigger Brains: Complex Brains for a Complex World": *http://humanorigins.si.edu/human-characteristics/brains* (September 2, 2015).

39   "Knuckle-walkers' wrist bones have several distinctive deatures. The radius (one of the bones in the forearm) and the wrist bones lock together during the weight bearing phase of knuckle-walking to form a solid supporting structure. Gorillas and chimps have these features, whereas humans do not." (Henry Gee, "These Fists were made for Walking," (March 23, 2000), *Nature News: www.nature.com/news/2000/000323/full/news000323-7.html* (September 1, 2015).

40   Brian G. Richmond & David S. Strait. "Lucy on the Ground With Knuckles," *Science News* 8 (April 2000): 235.

41   Credit: Kivell, T. L., & Schmitt, D. (2009). Proceedings of the National Academy of Sciences. 106, 14241–6.

42   Jack Stern & Randall Sussman. "The locomotor anatomy of *Australopithecus* afarensis." *American Journal of Physical Anthropology* 60 (1983): 279-317.

43   Brian G. Richmond & David S. Strait, "Evidence That Humans Evolved From a Knuckle-Walking Ancestor," *Nature*, 404 (6776) (March 23, 2000), 339–340, 382–385.

44   J.R. Napier, "Studies of the Hands of Living Primates," *Proceedings of the Zoological Society of London* 134 (1960): 647–657.

45   J.R. Napier & P.H. Napier. *The Natural History of the Primates* (MIT Press, 1985).

46   Elizabeth Mitchell, "Did Humans Walk Like Chimps Up the Evolutionary Tree?" (October 15, 2015), Answers in Genesis: *https://answersingenesis.org/human-evolution/did-humans-walk-like-chimps-up-the-evolutionary-tree/ (November 9, 2015).*
       See: Jerry Bergman, "Darwinism and the Nazi Race Holocaust," (November 1, 1999), Answers in Genesis: *https://answersingenesis.org/charles-darwin/racism/darwinism-and-the-nazi-race-holocaust/* (September 2, 2015);

47   Elizabeth Mitchell, "Did Humans Walk Like Chimps Up the Evolutionary Tree?" (October 15, 2015), Answers in Genesis: *https://answersingenesis.org/human-evolution/did-humans-walk-like-chimps-up-the-evolutionary-tree/ (November 9, 2015).*
   See: Jerry Bergman, "Darwinism and the Nazi Race Holocaust," (November 1, 1999), Answers in Genesis: *https://answersingenesis.org/charles-darwin/racism/darwinism-and-the-nazi-race-holocaust/* (September 2, 2015);

48   Credit: *https://answersingenesis.org/human-evolution/ape-man/flexi-feet-did-some-humans-fail-to-leave-them-in-the-trees/*

49   Friends of Washoe, "Chimpanzees in Zoo": *www.friendsofwashoe. org/learn/captive_chimps/zoos.html* (September 2, 2015).

50   W. Montagna. The skin of nonhuman primates. *American Zoologist* 12 (1972): 109-124.

51   Various sources will show minor differences in these comparisons. They are for example only.

52   Charles Darwin, *On the Origin of Species by Means of Natural Selection, or the Preservation of Favoured Races in the Struggle for Life* (London: John Murray, 1859).

53   Charles Darwin, *The Origin of Species by Means of Natural Selection, or the Preservation of Favoured Races in the Struggle for Life* (London: John Murray, 1872).

54   Marvin Lubenow, "Recovery of Neanderthal mtDNA: An Evaluation," *Creation Ex Nihilo Technical Journal* 12, 1 (1998): 89.

55   Erik Trinkaus, "Hard Times Among the Neandertals," *Natural History*, 87, 10 (December 1978: 58-63.

56   In a similar book (*In Man and the Lower Animals*, 1861), Huxley argued that humans, chimpanzees, and gorillas were more closely related to each other than any of them were to orangutans or gibbons.

57   Charles Darwin, *The Descent of Man and Selection in Relation to Sex* 2d ed. (London: John Murray, 1874): 178.

58   See: Jerry Bergman, "Darwinism and the Nazi Race Holocaust," (November 1, 1999), Answers in Genesis: *https://answersingenesis. org/charles-darwin/racism/darwinism-and-the-nazi-race-holocaust/* (September 2, 2015); David Klinghoffer, "Don't Doubt It: An important historic sidebar on Hitler and Darwin," (April 18, 2008). *National Review Online: www.discovery.org/a/4679* (see also bibliography list).

59 R.A. Hellman, "Evolution in American School Biology Books from the Late Nineteenth Century until the 1930s," *The American Biology Teacher* 27 (1968): 778-780.

60 Ernst Haeckel, *Anthropogenie, oder Entwicklungsgeschichte des Menschen* (Leipzig: Verlag von Wilhelm Engelmann, 1891).

61 Credit: *http://longstreet.typepad.com/thesciencebookstore/industrial_technological_art/page/7/*.

62 Carl C. Swisher III, Garniss H. Curtis, Roger Lewin, *Java Man: How Two Geologists Changed Our Understanding of Human Evolution* (Chicago: University of Chicago Press, 2000).

63 *The Living Age*, Volume 209 (Boston, MA: E. Littell & Company, 1896).

64 B. Theunissien, *Eugene Dubois and the Ape-Man from Java* (Norwell, MA: Kluwer Academic Publishers, 1989): 39.

65 Garniss Curtis, Carl Swisher, and Roger Lewin, *Java Man* (Abacus, London, 2000): 87.

66 James Perloff, *Tornado in a Junkyard: The Relentless Myth of Darwinism* (Burlington, MA: Refuge Books, 1999): 85.

67 Eugene DuBois, "On the Fossil Human Skulls Recently Discovered in Java and Pithecanthropus Erectus," *Man* 37 (January 1937): 4.

68 Credit: *http://jattwood.blogspot.com/2014/10/blog-3-java-man.html*

69 Credit: *http://kotawisataindonesia.com/museum-purbakala-sangiran-sragen/welcome-to-sangiran/*

70 Howard E. Wilson, "The Java Man (Pithecanthropus Erectus)" *Truth Magazine: www.truthmagazine.com/archives/volume2/TM002021.htm* (September 2, 2015).

71 *Science*, New Series, Vol. 57, June 15, 1923, Supplement 8.

72   *Science*, New Series, Vol. 58, Aug. 17, 1923, Supplement 8.

73   Marvin Lubenow, *Bones of Contention* (Grand Rapids, MI: Baker Books, 1992): 95.

74   *Science*, New Series, Volume 75, Supplement 11 (June 10, 1932) ("In the Smithsonian collection there are 32 American Indian skulls of small statured but otherwise apparently normal individuals ranging in capacity from 910 to 1,020 cc"). See also *1925 Science Supplement: Truth Magazine* (II:3, pp. 8-10, December 1957): *www.truthmagazine.com/archives/volume2/TM002021.htm* (September 2, 2015).

75   *1925 Science Supplement: Truth Magazine* (II:3, pp. 8-10, December 1957): *www.truthmagazine.com/archives/volume2/TM002021.htm* (September 2, 2015).

76   O.W. Richards, "The Present Content of Biology in the Secondary Schools," *School Science and Mathematics*, 23 (5) (1923): 409-414.

77   Pat Shipman, "On the Trail of the Piltdown Fraudsters," *New Scientist*, 128 (October 6, 1990): 52.

78   *Credit: http://up.botstudent.net/piltdown-man-new-york-times.jpg*

79   Lubenow, 1992, 42–43.

80   Arthur Keith, *The Antiquity of Man* (London: Williams & Norgate, 1915).

81   Arthur Keith, *The Antiquity of Man* (Philadelphia: J. B. Lippincott Company, 1928).

82   National Science Foundation, *Evolution of Evolution: Flash Special Report Timeline: www.nsf.gov/news/special_reports/darwin/textonly/timeline.jsp* (September 2, 2015).

83   Keith, 1915, 305.

84   Credit: *http://www.oakauctions.com/clarence-darrow-signed-%E2%80%9Cthe-antiquity-of-man%E2%80%9D-lot1674.aspx.*

85   *Nature* Volume 274, #4419 (July 10, 1954): 61-62.

86   Stephen Jay Gould, "Piltdown Revisited," *The Panda's Thumb* (New York: W.W. Norton and Company, 1982).

87   Credit: Science (May 20, 1927, p. 486) *http://bevets.com/piltdowng. htm*

88   Credit: Popular Science (October, 1931, p. 23) *http://bevets.com/ piltdowng.htm*

89   William K. Gregory, "Hesperopithecus Apparently Not an Ape nor a Man," *Science*, 66 (1720) (December 16, 1927): 579-581.

90   Ralph M. Wetzel, et al., "Catagonus, An 'Extinct' Peccary, Alive in Paraguay," *Science*, 189 (4200) (Aug. 1, 1975): 379.

91   Duane T. Gish, *Evolution: The Fossils Still Say NO!* (El Cajon, CA: Institute for Creation Research, 1995): 328.

92   Piltdown Man and Nebraska Man were mentioned in affidavits by "expert witnesses" Fay-Cooper Cole and Horatio Newman (professors at the University of Chicago), and Judge Raulston allowed their reports to be read into the court record on July 20, 1925.

93   Paul J. Wendel. *A History of Teaching Evolution in U. S. Schools: Insights from Teacher Surveys and Textbook Reviews* (Columbus, Ohio: Paper presented at the Mid-Western Educational Research Association, October 11-14, 2006).

94   J. I. Cretzinger, "An Analysis of Principles or Generalities Appearing in Biological Textbooks used in the Secondary Schools of the United States from 1800 to 1933," *Science Education* 25 (6) (1941): 310-313.

95   Gerald Skoog, "Topic of Evolution in Secondary School Biology Textbooks: 1900-1977," *Science Education* 63 (1979): 622.

96   The Topic of Evolution in Secondary Schools Revisited, Last updated February 15, 2010. *Textbook History: www.textbookhistory.com/ the-topic-of-evolution-in-secondary-schools-revisited/* (September 2, 2015).

97   Credit: *http://www.textbookhistory.com/the-topic-of-evolution-in-secondary-schools-revisited/*

98   Gerald Skoog, "Topic of Evolution in Secondary School Biology Textbooks: 1900-1977," *Science Education* 63 (1979): 622. See also: Paul J. Wendel. *A History of Teaching Evolution in U. S. Schools: Insights from Teacher Surveys and Textbook Reviews* (Columbus, Ohio: Paper presented at the Mid-Western Educational Research Association, October 11-14, 2006).

99   Credit: Life (May 21, 1951, p. 116) (*http://bevets.com/piltdowng. htm*)

100   Answers in Genesis, "Myth-making: The Power of the Image," (September 1, 1999): *https://answersingenesis.org/creationism/myth-making-the-power-of-the-image/* (September 2, 2015).

101   Clark F. Howell, *Early Man* (New York: Time Life Books, 1965): 41-45.

102   John Gurche (Sculptor), *National Geographic* 189 (3) (March 1996): 96-117.

103   Matthew Thomas (Podiatrist). Podiatry Arena (Blog): *www.podiatry-arena.com/podiatry-forum/showthread. php?t=68691&page=2* (September 2, 2015).

104   Chris Kirk, "Science the state of the Universe. Map: Publicly Funded Schools That Are Allowed to Teach Creationism," (January 26, 2014). Slate: *www.slate.com/articles/health_and_ science/science/2014/01/creationism_in_public_schools_mapped_ where_tax_money_supports_alternatives.html* (March 23, 2015).

105 America's Changing Religious Landscape, (May 12, 2015), *Pew Research Center: www.pewforum.org/2015/05/12/americas-changing-religious-landscape/* (September 2, 2015).

106 Lubenow, *Recovery of Neandertal mtDNA: An Evaluation*, 89.

107 Frank Newport, "In U.S., 46% Hold Creationist View of Human Origins: Highly Religious Americans most likely to believe in Creationism," (June 1, 2012), *Gallup: www.gallup.com/poll/155003/hold-creationist-view-human-origins.aspx* (March 23, 2015).

108 David Catchpoole and Tas Walker, "Charles Lyell's Hidden Agenda—to Free Science "from Moses" (August 19, 2009). *Creation.com: http://creation.com/charles-lyell-free-science-from-moses#endRef3* (March 23, 2015).

109 "Who are we?" (October 26, 2002), *New Scientist: www.newscientist.com/article/mg17623665-300-who-are-we/* (September 2, 2015).

110 "Return to the Planet of the Apes," *Nature*, Volume 412 (July 12, 2001): 131.

111 "The water people," *Science Digest*, Volume 90 (May 1982): 44.

112 Rinehart, Winston, & Holt, *Social Studies World History: Ancient Civilizations* (2006): 24-35.

113 PBS Evolution, "Finding Lucy": *www.pbs.org/wgbh/evolution/library/07/1/l_071_01.html* (September 2, 2015); National Geographic, "What was 'Lucy'? Fast Facts on an Early Human Ancestor" (September 20, 2006). *National Geographic News: http://news.nationalgeographic.com/news/2006/09/060920-lucy.html* (September 2, 2015).

114 Donald Johanson & Edgar Blake. *From Lucy to Language* (New York: Simon & Schuster, 1996).

115 NOVA, *In Search of Human Origins (Part I)* (Airdate: June 3, 1997): *http://www.pbs.org/wgbh/nova/transcripts/2106hum1.html* (September 2, 2015).

116   Time magazine reported in 1977 that Lucy had a tiny skull, a head like an ape, a braincase size the same as that of a chimp—450 cc. and "was surprisingly short legged" (*Time*, November 7, 1979, 68-69). See also: Smithsonian National Museum of Natural History, "Australopithecus afarensis": *http://humanorigins.si.edu/evidence/human-fossils/species/australopithecus-afarensis* (September 2, 2015).

117   Solly Zuckerman, *Beyond the Ivory Tower* (London: Taplinger Publishing Company, 1970): 78.

118   Solly Zuckerman and Charles Oxnard used mathematical studies of australopithecine fossils that revealed they did not have upright gait like humans. (Charles E. Oxnard, "The Place of the Australopithecines in Human Evolution: Grounds for Doubt?" *Nature* Volume 258 (December 4, 1975): 389–395; Solly Zuckerman, "Myth and Method in Anatomy," *Journal of the Royal College of Surgeons of Edinburgh*, Volume 11 (2) (1966): 87-114. Solly Zuckerman, *Beyond the Ivory Tower* (New York: Taplinger Publishing Company, 1971): 76-94; P. Shipman, "Those Ears Were Made For Walking,"*New Scientist* 143 (1994): 26–29; R. L. Susman & J.T. Susman, "The Locomotor Anatomy of *Australopithecus* Afarensis," *American Journal of Physical Anthropology* 60 (3) (1983): 279–317; and T. Eardsley, "These Feet Were Made for Walking – and?" *Scientific American* 273 (6) (1995): 19–20.

119   Charles Oxnard, *The Order of Man: A Biomathematical Anatomy of the Primates* (Yale University Press and Hong Kong University Press, 1984): 3.

120   Jack Stern & Randall L. Susman, "The Locomotor Anatomy of Australopithecus afarensis," *Journal of Physical Anthropology* 60 (1983): 280.

121   See Richmond & Strait, *Evidence That Humans Evolved From a Knuckle-Walking Ancestor*, 382-385.

122   Maggie Fox, "Man's Early Ancestors Were Knuckle Walkers," *San Diego Union Tribune* (Quest Section, March 29, 2000).

123   Richmond & Strait, *Evidence That Humans Evolved From a Knuckle-Walking Ancestor*, 382-385.

124   Donald Johanson & Maitland A. Edey, *Lucy: The Beginnings of Humankind* (London: Penguin, 1981): 358.

125   Ibid.

126   Richard Leakey & Roger Lewin, *Origins Reconsidered: In Search of What Makes us Human* (New York: Doubleday, 1992): 193-194.

127   Peter Schmid as quoted in Leakey and Lewin, *Origins Reconsidered*, 1992, 193-194.

128   Fred Spoor, Bernard Wood, Frans Zonneveld, "Implications of Early Hominid Labyrinthine Morphology for Evolution of Human Bipedal Locomotion," *Nature* 369 (June 23, 1994): 645-648.

129   M. Häusler & P. Schmid, "Comparison of the Pelves of Sts 14 and AL 288-1: Implications for Birth and Sexual Dimorphism in Australopithecines." *Journal of Human Evolution* 29 (1995): 363–383.

130   Alan Boyle, "Lucy or Brucey? It Can Be Tricky to Tell the Sex of Fossil Ancestors," *Science* (April 29, 2015).

131   Stern & Sussman, *The Locomotor Anatomy of Australopithecus afarensis*, 279-317.

132   Oxnard, *The Order of Man: A Biomathematical Anatomy of the Primates*, 3.

133   Roger Lewin, *Bones of Contention* (Chicago: University of Chicago Press, 1987): 164).

134   Wray Herbert, "Lucy's Uncommon Forbear," *Science News* 123 (February 5, 1983): 89.

135   Albert W. Mehlert, "Lucy—Evolution's Solitary Claim for an Ape/Man: Her Position is Slipping Away," *Creation Research Society*

Quarterly, 22 (3) (December, 1985): 145.

136 Lubenow, *Bones of Contention*, 179.

137 DeWitt Steele & Gregory Parker, *Science of the Physical Creation*, 2d ed. (Pensacola, FL: A Beka Book, 1996), 299.

138 "Before humans left Babel, it appears that apes had already spread over much of the Old World and had diversified into a large array of species... Paleontologists are still discovering species of post-Flood apes. If we are correct about post-Flood rocks, apes were at their highest point of diversity and were buried in local catastrophes just before humans spread out from Babel." Kurt Wise, "Lucy Was Buried First Babel Helps Explain the Sequence of Ape and Human Fossils," (August 20, 2008), *Answers in Genesis: https:// answersingenesis.org/human-evolution/lucy/lucy-was-buried-first/* (September 2, 2015).

139 The 'First Family' location Dr. Johanson refers to is within one mile of where Lucy was found (Donald Johanson as quoted in "Letters to Mr. Jim Lippard," *Institute of Human Origins* (August 8, 1989; May 30, 1990).

140 Not *A.* aferensus, but *Australopithecines* in general.

141 Donald Johanson, M. Taieb & Y. Coppens, "Pliocene Hominids from the Hadar Formation, Ethiopia (1973–1977): Stratigraphic, Chronologic and Paleoenvironmental Contexts, with Notes on Hominid Morphology and Systematics," *American Journal of Physical Anthropology* 57 (1982): 501–544.

142 Global Biodiversity Information Facility (GBIF), "What is GBIF": *http://www.gbif.org/whatisgbif* (September 2, 2015).

143 Anna K. Behrensmeyer, "Paleoenvironmental Context of the Pliocene A.L. 333 'First Family' Hominin Locality" (Hadar Formation, Ethiopia," *GSA Special Papers* (Volume 446, 2008): 203-214; Jay Quade and Jonathan G. Wynn, "The Geology of Early Humans in the Horn of Africa," (GSA Special Papers 446, 2008).

144   Donald Johanson, "Lucy, Thirty Years Later: An Expanded View of Australopithecus afarensis," *Journal of Anthropological Research* 60 (4) (Winter, 2004): 471-472.

145   Lauren E. Bohn, "Q&A: 'Lucy' Discoverer Donald C. Johanson," (March 4, 2009), Time: *http://content.time.com/time/health/article/0,8599,1882969,00.html* (September 2, 2015).

146   Darwin, *Origin of Species* (1859).

147   Darwin, *Origin of Species* (1872).

148   *Credit: http://humanorigins.si.edu/evidence/human-family-tree*

149   Smithsonian National Museum of Natural History, "*Homo habilis*": *http://humanorigins.si.edu/evidence/human-fossils/species/homo-habilis* (September 2, 2015).

150   Richard Leakey and Roger Lewin, *Origins Reconsidered* (New York: Doubleday, 1992): p 112. See also: Bernard Wood, "The age of australopithecines," *Nature* 372 (November 3, 1994): 31-32.

151   Lubenow, *Bones of Contention*, 300.

152   Richard Leakey and Roger Lewin, *Origins Reconsidered* (New York: Doubleday 1992): 112.

153   Fred Spoor, Bernard Wood, Frans Zonneveld, "Implications of Early Hominid Labyrinthine Morphology for Evolution of Human Bipedal Locomotion," *Nature* 369 (June 23, 1994): 645.

154   Ibid., 648.

155   Credit: *http://thefactofcreation.blogspot.com/2012/08/homo-habilis.html?m=1*

156   Bernard Wood, "The age of australopithecines," *Nature* 372 (November 3, 1994): 31-32.

157 Bernard Wood, "Human Evolution: Fifty years after *Homo habilis*," *Nature* 508 (April 2014): 31-33.

158 Holt, *Ancient Civilizations*, 30.

159 Bill Mehlert, "Homo Erectus to Modern Man: Evolution or Human Variability," (April 1, 1994): *Answers in Genesis: https:// answersingenesis.org/human-evolution/ape-man/homo-erectus-to-modern-man-evolution-or-human-variability/* (September 2, 2015).

160 Lubenow, *Bones of Contention: A Creationist Assessment of Human Fossils*, 115.

161 Ibid., 27-128.

162 Ibid., 115.

163 John Woodmorappe, "How Different Is the Cranial-vault Thickness of Homo Erectus from Modern Man?" *Creation.com: http://creation. com/how-different-is-the-cranial-vault-thickness-of-homo-erectus-from-modern-man* (September 2, 2015).

164 Ibid.

165 Lubenow, *Bones of Contention: A Creationist Assessment of Human Fossils*, 348-51.

166 Ibid.

167 Lubenow, *Bones of Contention: A Creationist Assessment of Human Fossils*, 130.

168 J. J. Hublin, "The Origin of Neandertals," *Proceedings of the National Academy of Sciences* 106 (38) (2009): 16022-7; K. Harvati, S.R. Frost, K.P. McNulty (2004), "Neanderthal Taxonomy Reconsidered: Implications of 3D Primate Models of Intra- and Interspecific Differences," Proceedings of the National Academy of Sciences, USA 101: 1147-1152.

169 David Menton & John UpChurch, "Who Were Cavemen? Finding a Home for Cavemen," (April 1, 2012), *Answers in Genesis: https://*

answersingenesis.org/human-evolution/cavemen/who-were-cavemen/ (September 2, 2015).

170   We are grateful to David V. Bassett, M.S. for his contributions to this section that were carried over from the first edition of *Creation V. Evolution*: What they Won't Tell You in Biology Class by Dan Biddle.

171   Marvin L. Lubenow, "Recovery of Neandertal mDNA: An Evaluation," *CEN Technical Journal*, 12 (1) (1998): 89.

172   Jack Cuozzo, *Buried Alive: The Truth about Neanderthal Man, Truths That Transform Action Sheet* (Radio Program, aired on March 14–15, 2000).

173   Lubenow, *Bones of Contention* (1992), 63.

174   Steele & Parker, *Science of the Physical Creation*, 301.

175   Lubenow, *Recovery of Neandertal mDNA: An Evaluation*, 89–90.

176   Jack Cuozzo, *Buried Alive: The Startling Truth About Neanderthal Man* (Green Forest, AZ: Master Books, 1998): 162, 163, 203.

177   Ibid.

178   Green, R. E. et al. A Draft Sequence of the Neandertal Genome. *Science* 328 (2010): 710–722.

179   Steele & Parker, *Science of the Physical Creation*, pp. 301–302.

180   Vance Ferrell, *The Evolution Cruncher* (Altamont, TN: Evolution Facts, Inc., 2001): 529.

181   Lubenow, 1992, p. 235.

182   Ian Taylor, "Fossil Man" Creation Moments Online: *http://www.creationmoments.com/content/fossil-man* (January 1, 2014).

183   Vance Ferrell, *The Evolution Cruncher* (Altamont, TN: Evolution Facts, Inc., 2001): 529.

184   Gerhard Meisenberg & William Simmons, *Principles of Medical Biochemistry* (New York: Mosby, 2006).

185   Susan Chavez Cameron and Susan Macias Wycoff, "The Destructive Nature of the Term 'Race': Growing Beyond a False Paradigm," *Journal of Counseling & Development*, Volume 76, no. 3 (Summer 1998): 277–285. The article cites information from L. Luca Cavalli-Sforza, Paolo Menozzi, and Alberto Piazza, *The History and Geography of Human Genes* (Princeton, NJ: Princeton University Press, 1994): 279.

186   Darwin, *Descent of Man, and Selection in Relation to Sex.*

187   Darwin, *Descent of Man, and Selection in Relation to Sex*, Volume 1, 201.

188   Stephen Jay Gould, *Ontogeny and Phylogeny* (Cambridge, MA: Harvard University Press, 1977): 127.

189   Darwin, *Descent of Man, and Selection in Relation to Sex*, Volume 2, 327.

190   Darwin, *Descent of Man, and Selection in Relation to Sex*, Volume 2, 328.

191   Taylor, 2009, 19.

192   Jonathan Silvertown (ed), *99% Ape: How Evolution Adds Up* (University of Chicago Press, 2009): 4.

193   Bagemihl, Bruce. 1999. *Biological Exuberance: Animal Homosexuality and Natural Diversity*. St. Martins Press. New York.

194   Many attempts have been made and all have failed. See Rossiianov, Kirill. 2002. "Beyond Species: Ii'ya Ivanov and His Experiments on Cross-Breeding Humans with Anthropoid Apes." *Science in Context*. 15(2):277-316.

195   See the U.S. Department of Health and Human Services (*https://optn.transplant.hrsa.gov/*) (February 1, 2016).

196   Various sources will show minor differences in these comparisons. These are for example only.

197   http://useast.ensembl.org/Homo_sapiens/Info/Annotation

198   Kakuo, S., Asaoka, K. and Ide, T. 1999. 'Human is a unique species among primates in terms of telomere length.' *Biochemistry Biophysics Research Communication.* 263:308-314

199   Archidiacono, N., Storlazzi, C.T., Spalluto, C., Ricco, A.S., Marzella, R., Rocchi, M. 1998. 'Evolution of chromosome Y in primates.' Chromosoma 107:241-246

200   "What about the Similarity Between Human and Chimp DNA?" *AnswersinGenesis.com: http://www.answersingenesis.org/articles/nab3/human- and-chimp-dna* (January 14, 2014).

201   R.J. Rummel, "Statistics of Democide: Genocide and Mass Murder Since 1900," *School of Law, University of Virginia* (1997).

202   Jerry Bergman. 2012. *Hitler and the Nazis Darwinian Worldview: How the Nazis Eugenic Crusade for a Superior Race Caused the Greatest Holocaust in World History.* Kitchener, Ontario, Canada: Joshua Press

203   J. Bergman & J. Tomkins, "Is the Human Genome Nearly Identical to Chimpanzee? A Reassessment of the Literature" *Journal of Creation* 26 (2012): 54–60.

204   Ibid.

205   J. Tomkins, "How Genomes are Sequenced and why it Matters: Implications for Studies in Comparative Genomics of Humans and Chimpanzees," *Answers Research Journal* 4 (2011): 81–88.

206   I. Ebersberger, D. Metzler, C. Schwarz, & S. Pääbo, "Genomewide Comparison of DNA Sequences between Humans and Chimpanzees," *American Journal of Human Genetics* 70 (2002): 1490–1497.

207   http://www.icr.org/article/6197/

208   Chimpanzee Sequencing and Analysis Consortium, "Initial Sequence of the Chimpanzee Genome and Comparison with the Human Genome," *Nature* 437 (2005): 69–87.

209   J. Tomkins, "Genome-Wide DNA Alignment Similarity (Identity) for 40,000 Chimpanzee DNA Sequences Queried against the Human Genome is 86–89%," *Answers Research Journal* 4 (2011): 233–241.

210   J. Prado-Martinez, et al. "Great Ape Genetic Diversity and Population History," *Nature* 499 (2013), 471–475.

211   J. Tomkins, & J. Bergman. "Genomic Monkey Business—Estimates of Nearly Identical Human-Chimp DNA Similarity Re-evaluated using Omitted Data," *Journal of Creation* 26 (2012), 94–100; J. Tomkins, "Comprehensive Analysis of Chimpanzee and Human Chromosomes Reveals Average DNA Similarity of 70%," *Answers Research Journal* 6 (2013): 63–69.

212   Nathaniel T. Jeanson, "Purpose, Progress, and Promise, Part 4," *Institute for Creation Research: http://www.icr.org/article/purpose-progress-promise-part-4* (September 2, 2015).

213   Tomkins & Bergman, 63-69.

214   Tomkins, 2011.

215   Tomkins & Bergman, 63-69.

216   Subsequent analyses revealed an anomaly in the BLASTN algorithm used for determining the 70% figure and the revised estimate (88%) has been included in this chapter. See: Jeffrey P. Tomkins, "Documented Anomaly in Recent Versions of the BLASTN Algorithm and a Complete Reanalysis of Chimpanzee and Human

Genome-Wide DNA Similarity Using Nucmer and LASTZ," (October 7, 2015), Answers in Genesis: *https://answersingenesis. org/genetics/dna-similarities/blastn-algorithm-anomaly/*

217   Tomkins, 2011.

218   E. Wijaya, M.C. Frith, P. Horton & K. Asai, "Finding Protein-coding Genes through Human Polymorphisms," *PloS one* 8 (2013).

219   New Genome Comparison Finds Chimps, Humans Very Similar at the DNA Level, 2005, national human genome research institute *http://www.genome.gov/15515096*

220   Christine Elsik. et al. The Genome Sequence of Taurine Cattle: A Window to Ruminant Biology and Evolution. *Science*. 324:522-528.

221   *http://www.eupedia.com/forum/threads/25335-Percentage-of-genetic-similarity-between-humans-and-animals*. Source is Pontius, Joan. Et. al. 2007. Initial Sequence and Comparative Analysis of the Cat Genome. *Genome Research*. 17:1675-1689.

222   http://www.genome.gov/10005835.

223   NIH/National Human Genome Research Institute. "Researchers Compare Chicken, Human Genomes: Analysis Of First Avian Genome Uncovers Differences Between Birds And Mammals." ScienceDaily. 10 December 2004.

224   M. J. Hangauer, I.W. Vaughn & M. T. McManus, "Pervasive Transcription of the Human Genome Produces Thousands of Previously Unidentified Long Intergenic Noncoding RNAs," *PLoS genetics* 9 (2013).

225   S. Djebali, et al. "Landscape of Transcription in Human Cells," *Nature* 489 (2012): 101–108.

226   M. D. Paraskevopoulou, et al. "DIANA-LncBase: Experimentally Verified and Computationally Predicted MicroRNA Targets on Long Non-coding RNAs," *Nucleic Acids Research* 41 (2013): 239–245.

227   J. J Yunis & O. Prakash, "The Origin of Man: A Chromosomal Pictorial Legacy," *Science* 215 (1982): 1525–1530.

228   J. W. Ijdo, A. Baldini, D.C. Ward, S. T. Reeders & R. A. Wells, "Origin of Human Chromosome 2: An Ancestral Telomere-telomere Fusion," *Proceedings of the National Academy of Sciences of the United States of America* 88 (1991): 9051–9055.

229   J. Bergman & J. Tomkins, "The Chromosome 2 Fusion Model of Human Evolution—Part 1: Re-evaluating the Evidence," *Journal of Creation* 25 (2011): 110–114.

230   J. Tomkins, "Alleged Human Chromosome 2 'Fusion Site' Encodes an Active DNA Binding Domain Inside a Complex and Highly Expressed Gene—Negating Fusion," *Answers Research Journal* 6 (2013): 367–375.

231   Y. Fan, E. Linardopoulou, C. Friedman, E. Williams & B.J. Trask, "Genomic Structure and Evolution of the Ancestral Chromosome Fusion Site in 2q13-2q14.1 and Paralogous Regions on other Human Chromosomes," *Genome Research* 12 (2002): 1651–1662; Y. Fan, T. Newman, E. Linardopoulou, & B.J. Trask, "Gene Content and Function of the Ancestral Chromosome Fusion Site in Human Chromosome 2q13-2q14.1 and Paralogous Regions," *Genome Research* 12 (2002): 1663–1672.

232   Y.Z. Wen, L. L. Zheng, L.H. Qu, F. J. Ayala & Z.R. Lun, Z. R, "Pseudogenes are not Pseudo Any More," *RNA Biology* 9 (2012): 27–32.

233   J. Tomkins, "The Human Beta-Globin Pseudogene Is Non-Variable and Functional," *Answers Research Journal* 6 (2013): 293–301.

234   M. Y. Lachapelle, & G. Drouin, "Inactivation Dates of the Human and Guinea Pig Vitamin C Genes," *Genetica* 139 (2011): 199–207.

235   J. Sanford, *Genetic Entropy and the Mystery of the Genome,* 3rd ed (FMS Publications, 2010).

236   J. Tomkins & J. Bergman, "Incomplete Lineage Sorting and Other 'Rogue' Data Fell the Tree of Life," *Journal of Creation* 27 (2013): 63–71.

237   Dan Biddle, *Creation V. Evolution: What They Won't Tell You in Biology Class* (Maitland, FL: Xulon Press); H. Morris, et al., *Creation Basics & Beyond: An In-Depth Look at Science, Origins, and Evolution* (Dallas, TX: Institute for Creation Research, 2013).

238   J. C. Sanford, *Genetic Entropy and the Mystery of the Genome*, 3rd ed. (Waterloo, NY: FMS Publications, 2008).

239   Ibid.

240   J. A. Tennessen, et al., "Evolution and Functional Impact of Rare Coding Variation from Deep Sequencing of Human Exomes," *Science* 337 (6090) (2012): 64-69; W. Fu, et al., "Analysis of 6,515 Exomes Reveals the Recent Origin of Most Human Protein-coding Variants," *Nature* 493 (7431) (2013): 216-220.

241   Tennessen, *Evolution and Functional Impact of Rare Coding Variation from Deep Sequencing of Human Exomes*, 64-69.

242   J. Sanford, J. Pamplin, & C. Rupe, "Genetic Entropy Recorded in the Bible?" (FMS Foundation. Posted on kolbecenter.org July 2014, accessed July 25, 2014).

243   N. T. Jeanson, "Recent, Functionally Diverse Origin for Mitochondrial Genes from ~2700 Metazoan Species," *Answers Research Journal* 6 (2013): 467-501.

244   T.J. Parsons, et al., "A High Observed Substitution Rate in the Human Mitochondrial DNA Control Region," *Nature Genetics* 15 (1997): 363-368.

245   A. Gibbons, "Calibrating the Mitochondrial Clock," *Science* 279 (1998): 28-29.

246   Acknowledgments for this chapter include Ted Siek, Clifford Lillo, John Woodmorappe, Jody Allen, Mary Ann Stuart, Bryce Gaudian, Don DeYoung, Paul Nesselroade, Stuart Burgess, and three anonymous reviewers.

247   S. Burgess, *The Design and Origin of Man: evidence for over design in the human being*. Surrey: DayOne Publications (2008).

248   S. Burgess, "Overdesign in the human being with a case study of facial expressions," *Journal of Creation*, 28 (1) (2014): 98-103.

249   J. Diamond, "Best Size and Number of Human Parts," *Natural History*, 103(6) (1994): 78.

250   H. Fisher, *The Sex Contract: The Evolution of Human Behavior* (New York: Morrow, 1982): 15.

251   S. Jones, *Darwin's Ghost: The Origin of Species updated* (New York: Random House, 2000): 98.

252   Burgess, *Overdesign in the human being with a case study of facial expressions*, 98-103.

253   Robert Ornstein, *Evolution of Consciousness: The Origins of the Way We Think* (New York, NY: Simon & Schuster, 1991): 40.

254   A. Alves, "Humanity's Place in Nature, 1863 – 1928: Horror, Curiosity and the Expeditions of Huxley, Wallace, Blavatsky and Lovecraft," *Theology and Science* 6 (1) (2008): 73-88.

255   A. R. Wallace, *The Limits of Natural Selection as Applied to Man* (chapter 10 in *Contributions to the theory of natural selection)* (London: Macmillan, 1870): 343.

256   Ibid., 359.

257   M. Shermer, "The Dangers of Keeping an Open Mind. Why Great Scientists Make Great Mistakes," *Scientific American* 309 (3) (2013): 92.

258   M. Shermer, *In Darwin's Shadow; the life and science of Alfred Russel Wallace* (New York: Oxford University Press, 2008): 163, 208; M. A.

Flannery, Alfred Russel Wallace's Theory of Intelligent Evolution (Riesel, TX: Erasmus Press, 2008).

259 Jones, *Darwin's Ghost: The Origin of Species updated*, 293.

260 S. Pinker, "The human mind," In J. M. Shephard, S. M. Kosslyn, & E. M. Hammonds (Eds.), *The Harvard Sampler: Liberal Education for the Twenty-first Century* (Cambridge, MA: Harvard University Press, 2011).

261 G. Simmons, *What Darwin Didn't Know: A Doctor Dissects the Theory of Evolution* (Eugene, OR: Harvest House Publishers, 2004): 89.

262 Simmons, *What Darwin Didn't Know: A Doctor Dissects the Theory of Evolution*, 89.

263 H. Ross, "Equipped for High-Tech Society," *Connections*, 6(3) (2004): 4.

264 Ibid.

265 Simmons, *What Darwin Didn't Know: A Doctor Dissects the Theory of Evolution*, 249.

266 Simmons, *What Darwin Didn't Know: A Doctor Dissects the Theory of Evolution*, 40.

267 D. Treffert, *Extraordinary People: Understanding "Idiot Savants"* (New York: Harper and Row, 1989).

268 S. Gould, *The Panda's Thumb: More Reflections in Natural History* (New York, NY: W. W. Norton, 1980): 55.

269 Wallace, *The Limits of Natural Selection as Applied to Man* (chapter 10 in *Contributions to the theory of natural selection*, 343, 356.

270 Robert K. Eberle, "If God Could Talk What Would he Say?" *Skeptic*, Volume 13, Issue 1 (March 2007): 78.

271 K. Rose, *The Body in Time*, (New York, NY: Wiley, 1988): 17-18.

272   Ibid., 18.

273   S. Smith, *The Great Mental Calculators: The Psychology, Methods, and Lives of Calculating Prodigies, Past and Present* (New York, NY: Columbia University Press, 1983): 237.

274   Ibid.

275   A. Snyder, "Savant-like numerosity skills revealed in normal people by magnetic pulses," *Perception* 35 (2006): 837.

276   F.V. Happé, "What Aspects of Autism Predispose to Talent? *Philosophical Transactions of the Royal Society of London. B. Biological Sciences* 364 (1522) (2009): 1369.

277   Treffert, *Extraordinary People: Understanding "Idiot Savants,"* 73-74.

278   N. Shanks, *God, the Devil and Darwin* (New York, NY: Oxford University Press, 2004): 194.

279   Rose, *The Body in Time,* 4.

280   Ibid.

281   P. Heaton & G. L. Wallace, "Annotation: the savant syndrome," *Journal of Child Psychology and Psychiatry* 45 (5) (2004): 899–911.

282   S. Rice, "Evolution of Intelligence," In *Encyclopedia of Evolution* (New York, NY: Facts on File, 2007): 209.

283   Smith, *The Great Mental Calculators: The Psychology, Methods, and Lives of Calculating Prodigies, Past and Present.*

284   Shanks, *God, the Devil and Darwin,* 78.

285   For example, see L. Obler & D. Fein, *The exceptional brain: Neuropsychology of talent and special abilities* (New York, NY: The Guilford Press, 1988).

286   T. Clark, "The Application of Savant and Splinter Skills in the Autistic Population Through Curriculum Design: a Longitudinal

Multiple-Replication Case Study," Ph.D. dissertation: University of New South Wales (2001).

287   W. Dembski, *Reflections on Human Origins,* Unpublished manuscript (2004): 1.

288   J. Hawking, *Music to Move the Stars* (New York, NY: McMillan, 2004): 200.

289   T. Wade, *The Home School Manual* (Auburn, CA: Gazelle, 1994): 283.

290   Dembski, *Reflections on Human Origins,* 2.

291   Dembski, *Reflections on Human Origins,* 4.

292   P. Senter, "Vestigial Skeletal Structures in Dinosaurs," *Journal of Zoology,* 280 (1) (January 2010): 60–71.

293   Thomas Heinze, *Creation vs. Evolution Handbook* (Grand Rapids, MI: Baker, 1973).

294   Isaac Asimov, *1959 Words of Science* (New York: Signet Reference Books, 1959), 30.

295   J. Bergman, "Are Wisdom Teeth (third molars) Vestiges of Human Evolution?" *CEN Tech Journal.* 12 (3) (1998): 297–304.

296   Charles Darwin, *The Descent of Man and Selection in Relation to Sex* (London: John Murray, 1871), 21.

297   Charles Darwin, *The Origin of Species* (New York: Modern Library, 1859), 346–350.

298   S. R. Scadding, "Do Vestigial Organs Provide Evidence for Evolution?" *Evolutionary Theory* 5 (1981): 173–176.

299   Robert Wiedersheim, *The Structure of Man: An Index to his Past History* (London: Macmillan, 1895, Translated by H. and M. Bernard).

300   David Starr Jordan & Vernon Lyman Kellogg, *Evolution and Animal Life* (New York: Appleton, 1908), 175.

301   Wiedersheim, 1895, p. 3.

302   Darwin, 1871, p. 29.

303   Cora A. Reno, *Evolution on Trial* (Chicago: Moody Press, 1970), 81.

304   Diane Newman, *The Urinary Incontinence Sourcebook* (Los Angeles, CA.: Lowell House, 1997), 13.

305   Warren Walker, *Functional Anatomy of the Vertebrates: An Evolutionary Perspective* (Philadelphia, PA: Saunders, 1987), 253.

306   Catherine Parker Anthony, *Textbook of Anatomy and Physiology*, 6th ed. (St. Louis, MO: Mosby, 1963), 411.

307   Anthony Smith, *The Body* (New York: Viking Penguin, 1986), 134.

308   Henry Gray, *Gray's Anatomy* (Philadelphia: Lea Febiger, 1966), 130.

309   Dorothy Allford, *Instant Creation—Not Evolution* (New York: Stein and Day, 1978), 42; Saul Weischnitzer, *Outline of Human Anatomy* (Baltimore, MD: University Park Press, 1978), 285.

310   J. D. Ratcliff, *Your Body and How it Works* (New York: Delacorte Press, 1975), 137.

311   Lawrence Galton, "All those Tonsil Operations: Useless? Dangerous?" *Parade* (May 2, 1976): 26.

312   Martin L. Gross, *The Doctors* (New York: Random House, 1966).

313   Jacob Stanley, Clarice Francone, & Walter Lossow, *Structure and Function in Man*, 5th ed. (Philadelphia: Saunders, 1982).

314   Alvin Eden, "When Should Tonsils and Adenoids be Removed?" *Family Weekly* (September 25, 1977): 24.

315 Syzmanowski as quoted in Dolores Katz, "Tonsillectomy: Boom or Boondoggle?" *The Detroit Free Press* (April 13, 1966).

316 Katz, 1972, p. 1-C.

317 N. J. Vianna, Petter Greenwald & U. N. Davies, "Tonsillectomy" In: *Medical World News* (September 10, 1973).

318 Katz, 1972.

319 Darwin, 1871, pp. 27–28.

320 Peter Raven & George Johnson, *Understanding Biology* (St. Louis: Times Mirror Mosby, 1988), 322.

321 Rebecca E. Fisher, "The Primate Appendix: A Reassessment," *The Anatomical Record*, 261 (2000): 228–236.

322 R. Randal Bollinger, Andrew S. Barbas, Errol L. Bush, Shu S. Lin and William Parker, "Biofilms in the Large Bowel Suggest an Apparent Function of the Human Vermiform Appendix," *Journal of Theoretical Biology*, 249 (4) (2007): 826–831; Thomas Morrison (ed.). *Human Physiology* (New York: Holt, Rinehart, and Winston, 1967).

323 Loren Martin, "What is the Function of the Human Appendix?" *Scientific American Online* (1999).

324 Thomas Judge & Gary R. Lichtenstein, "Is the Appendix a Vestigial Organ? Its Role in Ulcerative Colitis," *Gastroenterology*, 121 (3) (2001): 730–732.

325 Rod R. Seeley, Trent D. Stephens, & Philip Tate, *Anatomy and Physiology* (Boston: McGraw-Hill, 2003).

326 Ernst Haeckel, *The Evolution of Man: A Popular Exposition of the Principal Points of Human Ontogeny and Phylogeny* (New York: D. Appleton, 1879), 438.

327 Wiedersheim, 1895, p. 163.

328 O. Levy, G. Dai, C. Riedel, C.S. Ginter, E.M. Paul, A. N. Lebowitz & N. Carrasco, "Characterization of the thyroid Na+/I- symporter with an anti-COOH terminus antibody," *Proceedings from the National Academy of Science*, 94 (1997): 5568–5573.

329 Albert Maisel, "The useless glands that guard our health." *Reader's Digest* (November, 1966): 229–235.

330 John Clayton, "Vestigial Organs Continue to Diminish," *Focus on Truth*, 6 (6) (1983): 6–7.

331 Seeley, Stephens, & Tate, *Anatomy and Physiology* (McGraw-Hill Education, 2003), 778.

332 Maisel, 1966, p. 229.

333 Arthur Guyton, *Textbook of Medical Physiology* (Philadelphia: W. B. Saunders, 1966): 139.

334 Helen G. Durkin & Byron H. Waksman. "Thymus and Tolerance. Is Regulation the Major Function of the Thymus?" *Immunological Reviews*, 182 (2001): 33–57.

335 Durkin & Waksman, 2001, p. 49.

336 Benedict Seddon & Don Mason, "The Third Function of the Thymus," *Immunology Today*, 21 (2) (2000): 95–99.

337 Maisel, 1966.

338 Joel R. L. Ehrenkranz, "A Gland for all Seasons," *Natural History*, 92 (6) (1983): 18.

339 Stanley Yolles, "The Pineal Gland," *Today's Health*, 44 (3) (1966): 76–79.

340 David Blask, "Potential Role of the Pineal Gland in the Human Menstrual Cycle," Chapter 9 in *Changing Perspectives on Menopause*,

Edited by A. M. Voda (Austin: University of Texas Press, 1982), 124.

341   A. C. Greiner & S. C. Chan, "Melatonin Content of the Human Pineal Gland," *Science*, 199 (1978): 83–84.

342   Esther Greisheimer & Mary Wideman, *Physiology and Anatomy*, 9th ed. (Philadelphia: Lippincott, 1972).

343   Rosa M. Sainz, Juan C. Mayo, R.J. Reiter, D.X. Tan, and C. Rodriguez, "Apoptosis in Primary Lymphoid Organs with Aging," *Microscopy Research and Technique*, 62 (2003): 524–539.

344   Sharon Begley & William Cook, "The SAD Days of Winter," *Newsweek*, 155 (2) (January 14, 1985): 64.

345   Sainz, et al., 2003.

346   G.J. Maestroni, A. Conti, & P. Lisson, "Colony-stimulating activity and hematopoietic rescue from cancer chemotherapy compounds are induced by melatonin via endogenous interleukin," *Cancer Research*, 54 (1994): 4740-4743.

347   B.D. Jankovic, K. Isakovic, S. Petrovic, "Effect of Pinealectomy on Immune Reactions in the Rat," *Immunology*, 18 (1) (1970): 1–6.

348   Lennert Wetterberg, Edward Geller, & Arthur Yuwiler, "Harderian Gland: An Extraretinal Photoreceptor Influencing the Pineal Gland in Neonatal Rats?" *Science*, 167 (1970): 884–885.

349   Ehrenkranz, 1983, p. 18.

350   Philip Stibbe, "A Comparative Study of the Nictitating Membrane of Birds and Mammals," *Journal of Anatomy*, 163 (1928): 159–176.

351   Darwin, 1871, p. 23.

352   Henry Drummond, *The Ascent of Man* (New York: James Potts and Co., 1903).

353   Richard Snell & Michael Lemp, *Clinical Anatomy of the Eye* (Boston:

Blackwell Scientific Pub, 1997), 93.

354 Eugene Wolff (Revised by Robert Warwick), *Anatomy of the Eye and Orbit* 7th ed. (Philadelphia: W B. Saunders, 1976), 221.

355 John King, Personal communication, Dr. King is a professor of ophthalmology at The Ohio State School of Medicine and an authority on the eye (October 18, 1979).

356 E. P. Stibbe, "A Comparative Study of the Nictitating Membrane of Birds and Mammals," *Journal of Anatomy* 62 (1928): 159–176.

357 Wiedersheim, 1895.

358 D. Peck, "A Proposed Mechanoreceptor Role for the Small Redundant Muscles which Act in Parallel with Large Prime movers" in P. Hinick, T. Soukup, R. Vejsada, & J. Zelena's (eds.) *Mechanoreceptors: Development, Structure and Function* (New York: Plenum Press, 1988), 377–382.

359 David N. Menton, "The Plantaris and the Question of Vestigial Muscles in Man," *CEN Technical Journal*, 14 (2) (2000): 50–53.

360 Herbert DeVries, *Physiology of Exercise for Physical Education and Athletics* (Dubuque, IA: William C. Brown, 1980), 16–18.

361 Newport, *In U.S., 46% Hold Creationist View of Human Origins: Highly religious Americans most likely to believe in Creationism.*

362 D.M. Lloyd-Jones, *What is an Evangelical?* (Edinburgh, Scotland: Banner of Truth Trust, 1992): 75.

CPSIA information can be obtained
at www.ICGtesting.com
Printed in the USA
FSOW01n0602170616
21645FS